Creative Ways of Knowing in Engineering

Diana Bairaktarova • Michele Eodice
Editors

Creative Ways of Knowing in Engineering

Springer

Editors
Diana Bairaktarova
Department of Engineering Education
Virginia Tech
Blacksburg, VA, USA

Michele Eodice
University of Oklahoma
Norman, Oklahoma, USA

ISBN 978-3-319-49351-0 ISBN 978-3-319-49352-7 (eBook)
DOI 10.1007/978-3-319-49352-7

Library of Congress Control Number: 2016961694

Printed on acid-free paper

This Springer imprint is published by Springer Nature
The registered company is Springer International Publishing AG
The registered company address is: Gewerbestrasse 11, 6330 Cham, Switzerland

Foreword

At their most creative, engineers envision sustainable ways to provide communities with the essentials for life—water, food, shelter, energy, and security—as well as the necessary resources and support services like transport, physical and electronic infrastructure, and healthcare that enable well-being. Through imaginative design and inventive ways of making things, engineers shape our built environment through the provision of new products, processes, and systems. Working in interdisciplinary contexts, they help create objects of both utility and beauty, enabling individuals and multicultural communities to meet their diverse needs and to achieve higher personal and collective aspirations.

This timely book explores a variety of innovative pedagogical strategies for engaging students in creativity-enhancing ways of experiencing what means to engineer. Being serenaded in song about the virtues of thermodynamics or composing a poem about a dry scientific law or making a fun video about an engineering concept or being part of a conductor-less orchestra is not what first comes to mind when we think of a day in the life of an engineering student. Yet these are amongst the eclectic range of learning performances that are discussed in this book.

Engineering is not a narrow technical pursuit. It is both an art and a science, drawing on many types of knowledge and divergent ways of understanding the world. The successful application of technical knowledge depends critically on enabling knowledge about the human condition; knowledge from the social and behavioral sciences, the liberal arts, and humanities. Engineering is a profoundly human endeavor. Success as an engineer depends vitally upon being self-aware, emotionally intelligent, empathetic, an active listener and a nuanced communicator with diverse groups, persuasive both orally and in all manner of written styles, trustworthy and collaborative, and able to perform in structured teams as well as ad hoc groups that emerge in the course of a project. These essential professional abilities are sometimes referred to as "soft skills," but this is a misnomer. While engineering has a reputation for being a difficult discipline to master due to the emphasis on the "hard" sciences and mathematics, the truth is these knowledge domains are relatively easy to distil and transmit as compared to the messy process of formation of professional knowledge and skills and the development of character, judgement, insight, and ultimately wisdom.

This book reminds us that the most enlightened engineering education has always fostered an appreciation in our students that to engineer is to be a creative agent engaged in the most pressing local, national, and global conversations of the day. In the enthusiastic embrace of the engineering sciences half a century ago, much of engineering education lost sight of the nontechnical dimension of the profession. In this process even the defining art of the engineer, design, almost vanished from curricula. It is pleasing to see that the balance between the art and the science in engineering is being restored. A critical dimension of this restoration is having a deep appreciation of the professional skills being integral to the education of an engineer.

This process of rebalancing implies that engineering educators, who are themselves engineers, must engage in meaningful conversations with their peers in the arts and humanities and indeed education on how best to accomplish a more holistic engineering education experience. These conversations must be predicated on an openness to discovering there are other ways of knowing, a willingness to learn about alternative modes of inquiring into the world and a better understanding of the epistemological foundations of our own discipline. The critical need for such conversations should not be underestimated. Engineering academics have an unfortunate tendency to be all wise and assume they can master all there is to be known about professional skills and how best to incorporate the development of these skills into courses and curricula. On the contrary, the creation of learning experiences that blend technical and professional knowledge and skills in new ways needs to be a collaborative effort, an interdisciplinary partnership based on mutual respect and appreciation for what different nonengineering perspectives can bring to the table.

Such an interdisciplinary collaboration is exemplified by the editors of this book, Diana Bairaktarova and Michele Eodice. I have long argued that we need radical new approaches in engineering courses and curricula to re-establish a more appropriate balance between technical knowledge, know-how and skills, and the formation of professional abilities. The imaginative design of co-created, interdisciplinary courses must be founded upon relationships of trust and respect between academics from vastly different intellectual traditions. This co-creation may also involve students from the different disciplines. Building such relationships takes time and a sustained commitment; they cannot be rushed or planned out on a rigid timetable. Often serendipity plays a major part; being open to unexpected opportunities that arise or simply being mindful, being present in the moment, is an essential ingredient. This work requires initiative and resourcefulness, taking professional risks and showing personal courage, a strong determination to succeed, and persistence in the face of opposition, obstacles, and setbacks. In short it requires grit.

Formal education tends to privilege particular types of knowledge and empirical ways of knowing. Our educational system, based as it is on a nineteenth-century industrial conception of production, has a tendency to suppress imagination. This can be very discouraging for those who learn differently or use ways of knowing that do not fit the prevailing analytical paradigm or conventional assessment regime. Yet the apparent misfits in the current educational system may be the very people we need to take up a career in engineering, in order to truly diversify the profession. For engineering to reach its full potential, we need to attract and retain a broader range of people, people who look at the world in different ways and people who deviate from our current ways of thinking.

Engineering needs to be more inclusive of individuals from groups in society that are currently underrepresented in its ranks but who possess the abilities that our current system of engineering education values and is designed for. But more than that we must diversify our student body in terms of those from currently represented and underrepresented groups who bring other types of skill and knowledge, contrarians who exhibit nontraditional attributes that will enrich engineering education and transform professional practice through enlarging what it means to engineer. Many of the pedagogical experiments described in the book point to ways we might open the doors to a broader participation in engineering in both these ways.

Engineering is a global profession. Its practice may involve engineers and other disciplines drawn from many national and ethnic backgrounds, educated in different countries yet its impact is felt locally, in a particular socio-cultural, economic, and political context. Likewise engineering education is a major global export, where tens of thousands of students from many nations are educated as engineers in countries other than their own and in a language that is not their mother tongue. Engineering courses and curricula are exported to countries that may have quite different cultural precepts, philosophical understandings, and social mores to those where the learning was designed. There is no universal model for how engineers might be best educated; one size does not fit all. As one of the chapters in the book illustrates, we need to be thoughtful and employ cultural sensitivity in translating educational practices.

The most prized possession of any professional is their integrity. In your dealings as an engineer, who you are as person is far more important than what you know or what you can do. Accordingly, engineering educators have a solemn responsibility to their students and to the communities they serve to help their students to know themselves, to understand the obligations of being professional, to explore their innate moral compass as well as become knowledgeable about various ethical precepts and frameworks, and to provide them with useful tools for critical self-reflection to guide them throughout their career. In earlier eras, when most engineering academics had experience working as an engineer and design was still a central plank of engineering education, communicating the import of what it means to be a professional and how you might prepare to accept sometimes awesome responsibilities was an osmotic process that took place over an extended time. The wisdom of the years was imparted over the student's drawing board through numerous exchanges with instructors based on personal accounts, "war stories," and discussions about notable engineering failures. A variety of ethical questions were explored through these exchanges, and the moral dilemmas of engineering practice were ever present in the shadow of the then recent events of WW II and the tensions of Cold War. Many professors modeled what it was difficult to be a "reflective practitioner," even though reflection was not formally part of the curriculum.

We live in quite different era. The underlying educational assumptions have changed, the curriculum is very different, most academics have not had the benefit of practicing engineering, the classroom experience is being transformed, the sheer scale of the engineering education enterprise is grown considerably, and every student has instant access at the fingertips to global informational resources on a previously unimaginable scale. So educators must now be more intentional about incorporating instruction and learning experiences that actively foster identity development, to expose students to the legal and ethical dimensions of engineering and to develop the ability to critically reflect. This book provides numerous ideas and innovative approaches as to how we might enhance all three of these elements of professional formation appropriate to contemporary conditions of engineering education. Some of these approaches are quite radical, even confronting. All the more reason to engage with them and to challenge our implicit assumptions about what learning to engineer might look like.

The book also highlights the need to develop critical thinking in engineering students. This is the foundation for developing effective decision making under uncertainty and with incomplete information, essential attributes for a professional engineer. It helps students become more comfortable living with ambiguity rather than relying on the false certainty and illusory confidence afforded by neat and tidy closed form problems with simple solutions. Engineers work on "wicked problems" and are called upon to make life and death judgment calls. In the academy, the development of critical thinking skills has historically been seen as the province of the liberal arts and humanities where there are no simple answers to complex questions and positions are arrived at on the basis of reasoned arguments. This way of thinking and working makes many engineers, engineering educators, and engineering students at best uncomfortable and skeptical and at worst, incredulous, cynical, and even dismissive. This is why universities and colleges with engineering schools must also have a vibrant liberal arts community, valued on its own intellectual terms. The sorts of deep interdisciplinary insight needed to foster authentic critical thinking in engineering programs depend profoundly upon founts of disciplinary excellence beyond engineering and respectful boundary crossing in both directions by academics of good will.

The coeditors of this book, Diana Bairaktarova and Michele Eodice, are such boundary agents. They have developed a strong professional working relationship and rapport based on a deep appreciation for what the other can bring to the conversation. This relationship led to the creation of this eclectic collection of papers written by fellow travelers on this journey of exploration of *Creative Ways of Knowing in Engineering*. My hope is that the example set by these pioneers will encourage other engineering educators to reach out and partner with colleagues on the other side of campus in the liberal arts and humanities, education, and the other social and behavioral sciences to explore innovative and fun ways of stimulating the artist in their engineering students.

Swinburne University of Technology,
Hawthorn, VIC, Australia
Purdue University, West Lafayette, IN, USA

David F. Radcliffe

Acknowledgements

This work is the outcome of the combined efforts of many people—most especially the creative students and teachers who talk at length and with great passion about their creative ways of knowing in engineering. This book has, in a manner of speaking, been less than a year in the making, and there are more people to thank for the ideas reflected here than can possibly be named. Regardless, we must make an attempt.

We thank David Radcliffe, whose interdisciplinary work through the years as well as the work of many other leaders in the STEM education field has paved the way for current efforts to include the arts in STEM.

In describing her own journey, Diana wants to mention especially the influence of her Professor, Dr. William Graziano, whose mentorship has been uplifting and inspiring. Life has strange way of changing the paths we take and sometimes introduces us to people we feel we have always known. Professor Graziano, who has inspired Diana to care about her students and to creatively search for new knowledge, embodies and exemplifies an academic soul very similar to Diana's father's. Through these influences, Diana built confidence that she can motivate and inspire her students to strive to do their best work. We dedicate this book to Dr. William Graziano, and to all the creative teachers across this country and the world, whose innovative and cross disciplinary work encourages creative thinking and work in education.

Diana owes a bigger debt of gratitude than she can ever express to her friend, colleague, and coeditor, Michele Eodice. Michele was a great inspiration for this edited collection and a source of many flourishing ideas of creative ways of knowing in engineering.

Cheryl Cohen was heroic in her editorial assistance with an early draft of the first chapter of this book, and Desen Ozkan was unfailingly supportive and helpful in every imaginable way. We hope Desen, a future engineering educator and researcher, will continue to passionately enact a shift back to artistry in engineering education in order to successfully bridge the humanities and engineering.

Diana thanks her husband Michael and her son Nikola for being the inspiration and constant support for her work.

Michele would like to thank Diana Bairaktarova for her friendship and for the opportunity to learn more about teaching from an excellent teacher. When Diana opened her classroom and curriculum to Michele and to others, they had a chance to see and share the "goodness" of students, and their excitement for learning as a reminder of why we work in higher education. Thanks also go to Kami Day for her support, editorial and otherwise.

Contents

About the Authors

Diana Bairaktarova is an assistant professor of Engineering Education at Virginia Tech and the Director of the Creativity Inspiration Engineering Design Aptitudes and Abilities (CIEDAA) Lab. She holds an M.S. degree in Mechanical Engineering, an M.B.A., and a Ph.D. in Engineering Education from Purdue University. Bairaktarova's ongoing research interests span from engineering to psychology to learning sciences, as she uncovers how individual performance and professional decisions are influenced by aptitudes and abilities, personal interests, and manipulation of physical and virtual objects.

Diana joined the Department of Engineering Education in the fall of 2015 after being an assistant professor of engineering practice at the University of Oklahoma's College of Engineering. She taught several fundamental and engineering design courses in the School of Aerospace and Mechanical Engineering, where the design of artifacts was addressed from a multidisciplinary perspective that includes opportunity determination through inspiration, ideation, and implementation using a design thinking framework. She is a past recipient of the University of Oklahoma Presidential Dream Course Award for her course "User-Experience Design: From Renaissance Engineering to Design for Innovation." At the University of Oklahoma, Diana was also the proud faculty advisor of the Sooner-Off Road student-led team.

Diana has over 15 years of experience working as a Design and Manufacturing Engineer. By providing applications of problem- and project-based learning in the exploration of new designs that stimulate creativity, Dr. Bairaktarova aims to prepare her students with innovative thinking and a desire to acquire new skills and knowledge, preparing them to face rapidly changing technologies. Ranging from the exploration of humanly made objects to the education of her students, she strives to enhance her students' ability to explore and express their creativity, discover their own potential talent, and ultimately bring their ideas to fruition.

Michele Eodice is the Associate Provost for Academic Engagement and Director of the OU Writing Center at the University of Oklahoma. She earned a Ph.D. in English, writing her dissertation on coauthoring and collaborative writing in the classroom. Eodice's ongoing research interests include coauthoring, collaborative writing, adult and higher education, developing faculty writing at universities, and student engagement and learning through writing practices.

From 1998 to 2006, Eodice was the founding director of the writing center at the University of Kansas. Currently she is a professor of writing at the University of Oklahoma and as a program director and associate provost is involved with writing across the curriculum and other initiatives, such as academic service learning and community engagement.

At the University of Oklahoma, Eodice also holds affiliate faculty appointments with the Department of English and with the Adult and Higher Education program in the Jeannine Rainbolt College of Education.

For many years Eodice was in leadership roles within the International Writing Centers Association, serving as president from 2007 to 2009. During that timeframe she traveled to several countries as a consultant in developing writing centers abroad. For 6 years she was a cochair and facilitator for the IWCA Summer Institute for Writing Center Directors and Professionals.

Eodice has been a director of a writing center and a leader in the field of writing centers for 20 years; she currently serves as editor of *The Writing Center Journal*, the primary research journal of the International Writing Centers Association (writingcenterjournal.org).

Among her publications, two books are the products of important collaborations, *(First Person)2: A Study of Co-Authoring in the Academy* (2001), written with Kami Day, and *The Everyday Writing Center: A Community of Practice* (2007), written with Anne Ellen Geller, Frankie Condon, Meg Carroll, and Elizabeth H. Boquet.

Eodice works extensively with faculty and graduate student writers and facilitates writing groups, camps, and retreats across the country. With Anne Ellen Geller as coeditor, she published *Working with Faculty Writers* (2013), a book that details the range of national best practices in programmatic support for faculty writers.

Several contributions to collections have expanded a career-long theme of work that combines understanding writing practices and students' learning of writing with collaboration and coauthoring. One chapter, "Creativity in the Writing Center," written with Elizabeth Boquet, appears in a 2009 award winning collection, *Creative Approaches to Writing Center Work.* Her interest in creativity extends to other fields as well, including engineering.

Also, with co-researchers Anne Ellen Geller and Neal Lerner, she published *The Meaningful Writing Project: Learning, Teaching, and Writing in Higher Education.* This study of student writing experiences and faculty connections to writing development across the disciplines continues to invite participation (see: meaningfulwritingproject.net).

Currently at OU she focuses on supporting graduate student and faculty writers and forwarding the goals of the Writing Enriched Curriculum program.

Contributors

Diana Bairaktarova Department of Engineering Education, Virginia Tech, Blacksburg, VA, USA

Lindy Cranwell Department of Civil and Environmental Engineering, Virginia Tech, Blacksburg, VA, USA

Heather Marcelle Crickenberger University of North Carolina at Charlotte, Charlotte, NC, USA

Diana Dabby Franklin W. Olin College of Engineering, Needham, MA, USA

Gillian Epstein Franklin W. Olin College of Engineering, Needham, MA, USA

Mary Faure Department of Engineering Education, The Ohio State University, Columbus, OH, USA

Christopher Gewirtz Department of Engineering Education, Virginia Tech, Blacksburg, VA, USA

Cassandra Groen Department of Engineering Education, Virginia Tech, Blacksburg, VA, USA

Lynn Hall Department of Engineering Education, The Ohio State University, Columbus, OH, USA

Jennifer L. Herman Department of Engineering Education, The Ohio State University, Columbus, OH, USA

Deborah Kuzawa Department of Engineering Education, The Ohio State University, Columbus, OH, USA

Lisa D. McNair Department of Engineering Education, Virginia Tech, Blacksburg, VA, USA

Kevin C. Moore University of California, Santa Barbara, Santa Barbara, CA, USA

Marie C. Paretti Department of Engineering Education, Virginia Tech, Blacksburg, VA, USA

Lynne Ronesi American University of Sharjah, Sharjah, United Arab Emirates

Adetoun Taiwo Department of Engineering Education, Virginia Tech, Blacksburg, VA, USA

Leah Wahlin Department of Engineering Education, The Ohio State University, Columbus, OH, USA

Rabih Younes Department of Electrical and Computer Engineering, Virginia Tech, Blacksburg, VA, USA

Yevgeniya V. Zastavker Franklin W. Olin College of Engineering, Needham, MA, USA

The New Renaissance Artificers: Harnessing the Power of Creativity in the Engineering Classroom

Diana Bairaktarova

Engineering of the Future

The 2020 initiative proposed by the National Academy of Engineers called for creativity and interdisciplinary thinking for our engineering graduates. These skills and abilities were identified as critical for this new century as companies seek graduates who possess an increasingly broader set of skills—what the National Academy of Engineers termed *renaissance engineers*. These are technically proficient engineering graduates whose education extends beyond conventional engineering education in both technical and nontechnical ways, including the development of creative skills (National Academy of Engineering, 2004).

In the fall of 2016, we started teaching the cohort that will graduate in 2020. Are we ready to graduate the first cohort of renaissance engineers? Changes in engineering education are taking place across universities. A new discipline, called engineering education, and many new schools of engineering education were created. Our best programs now foster experiential learning and encourage multidisciplinary teams, offering learning experiences different from those in a traditional engineering curriculum. However, these innovative approaches are not happening in all engineering schools and not all at the same rate (Ottino & Morson, 2016). ArtScience Labs are flourishing around the world, acting as a stimulating catalyst for innovation by fusing the creative processes of artists and scientists alike. Regardless, David Radcliffe states "*these are exciting developments, but we need to foster organic interdisciplinary collaborations between scholars and practitioners across all STEM disciplines and the liberal arts*" (Radcliffe, 2015).

We are presented with the challenge of creating an educational space that can successfully combine art, technology, and science as a united phenomenon to ulti-

D. Bairaktarova (✉)
Department of Engineering Education, Virginia Tech, 865 Prices Fork,
Blacksburg, VA 24060, USA
e-mail: dibairak@vt.edu

D. Bairaktarova, M. Eodice (eds.), *Creative Ways of Knowing in Engineering*,
DOI 10.1007/978-3-319-49352-7_1

mately transform our classrooms into workshops and studios bursting with activity. To prepare the renaissance engineers of the future we look to the past, where renaissance artificers embraced art, technology, and science. This phenomenon was not isolated, but rather the supremely creative culmination of a long process. Throughout this chapter, I invite you to take a journey with me to the past.

Renaissance Engineering

The period of greatest flowering of modern Western thought is the Italian renaissance. The Italian renaissance was characterized by paradigm shifts in both art and science. Common across the disciplines of art and science was an emphasis on observation of the natural world. Inventiveness reflected great interest in nature, figures were three dimensional, and shadows and lighting were heavily considered. These innovations emphasized the virtues of intellectual freedom and individual expression where the instinct of curiosity was vigorously cultivated (King, 2003).

Drawing and sketching were tools that enabled (or developed) the practice of observing nature. Thinkers/artists recorded their observations in drawings that allowed them to test and refine their ideas about the natural world. For example, Galileo Galilei used his drawing talent and refined knowledge of perspective to make watercolor images of the moon. His models of the moon (informed by his mastery of perspective knowledge) allowed him to refine Copernican theories about the rotation of the earth around the sun. Galileo continued to use his mastery of perspective in drawing to translate three dimensions into two dimensions by manipulating the light and shadows of geometric forms.

In his book *Renaissance Engineers*, Paulo Galluzzi, director of the Museum of History of Science in Florence, provides us with the opportunity to reassess the Renaissance. His book examines the work of four great artist-engineers of the Italian Renaissance: Filippo Brunelleschi, Taccola, Francesco di Giorgio, and Leonardo DaVinci, and features large-scale reproductions of their drawings and designs of "mechanical marvels," which were the most significant technological achievements of that time. Among these engineering artifacts featured by Galluzzi are the dome of the Florence Cathedral (engineered by Filippo Brunelleschi) and engineering manuscripts that illustrate harnessing water, conducting warfare, potential machines, and the energy source for machines (created by Taccola and Francesco di Giorgio).

In the last section of the book, Galluzzi keeps a special place for DaVinci's work on machines and mechanisms: the anatomy of machines, the human body as a wonderful machine, his lost robot, the body–earth analogy, and the machine—building. DaVinci's most innovative contribution was that he was the first one to look at machines as a system assembled from many individual parts, which he referred to as "elements of machines" (Galluzzi, 1996). He analyzed the performance and characteristics of these elements by measuring force and motion.

DaVinci applied the same systematic method to study the human body and the internal organs he regarded as highly sophisticated mechanical devices. DaVinci was not only an engineer, mathematician, anatomist, and inventor but he was also painter, sculptor, musician, and writer. Because of his many talents, his vision of the anatomy of machines and of humans was enshrined in a series of masterly drawings that mark the birth of modern scientific illustration (Galluzzi, 1996).

Galluzzi refers to the multidisciplinary work of DaVinci and other renaissance creators as the universal intellectual experience—embracing art, technology, and science (Galluzzi, 1996). Galluzi believed that the multidisciplinary talents of Renaissance artists were not an isolated phenomenon, but rather part of a long process that renewed technical knowledge and persisted throughout the Renaissance.

He referred to these skilled artist-scientists as "artificers," who were "*instrumental in defining and subsequently winning recognition for a new breed of professional the artist-engineer- architect-author*" (Galluzzi, p. 4).

When examining the work of those expressively talented people, there is a general consensus that the Renaissance culture displays a fusion of the creative processes of artists and scientists. Is it the time period that naturally engenders this universal intellectual experience—embracing art, technology, and science? Or is it the artificers' expression of individuality and curiosity, looking for the truth in nature? One may argue that the DNA of those talented and advanced-for-their-time people is very close to the Innovator's DNA of the twenty-first century.

The Innovator's DNA

In "The Innovator's DNA," Dyer, Gregersen, and Christensen (2011) build on what we know about paradigm shifts to characterize the behaviors of the world's best innovators. The authors conducted research on 500 inventors compared to close to five thousand executives and identified five inventive skills that distinguish innovative leaders from ordinary leaders: associating, questioning, observing, networking, and experimenting. Dyer and colleagues argue that innovators use associative thinking to synthesize and make sense of discovery by making connections across seemingly unrelated questions, problems, or ideas. This phenomenon is described by Frank Johanssen, a writer, entrepreneur, and scientist, as "the Medici effect." Johanssen, with an interdisciplinary background himself, refers to the fusion of the creative processes of artists and scientists in Florence to the time when the Medici family brought together creators from a wide range of disciplines—sculptors, scientist, poets, philosophers, painters, and architects (Johansson, 2013). The message Johanssen has been delivering through his books, leadership, and public speaking is that breakthrough innovations happen at the intersection of diverse disciplines, cultures, and ideas and not being afraid to be creative and thinking outside of the box.

Early research on creative abilities provides empirical evidence that thinking outside of the box is not genetic but rather malleable. Reznikoff, Domino, Bridges, and Honeymon (1973) completed a comprehensive study studying creative abilities in 117 pairs of identical and fraternal twins. Their findings revealed that only about 30 % of the performance of identical twins on a battery of ten creativity tests could be attributed to genetics. In contrast, 85 % of the twins' performance on general intelligence (IQ) tests was attributed to genetics. This study supports the hypothesis that general intelligence (as measured by IQ tests) is genetically endowed but creativity is not (Reznikoff et al., 1973). Consequently, we can see that creativity is a learned practice and we need to shift our educational practices accordingly.

Teaching and Learning Creativity

Recently, the call for innovation has garnered a greater interest in creative aptitude. Increased efforts are encouraging universities to produce graduates with creative thinking skills, who are flexible, adaptable, and able to solve problems in order to face the challenges of the twenty-first century (Grainger, Barnes, & Scoffham, 2004). Despite this increased insistence that creativity is a 'good thing,' it is poorly understood and difficult to define (Coate & Boulos, 2012). Some scholars question its role in education and its relevance across cultural and societal contexts (Craft, 2003). Craft argues that there are paradigm shifts in the concept of creativity from extraordinary creativity to ordinary creativity. Furthermore, creativity is now understood as a culturally specific phenomenon as opposed to a universal quality. In "*The Limits to Creativity in Education: Dilemmas for the Educator*" Craft contends that creativity is not a universal concept; the author poses a set of dilemmas for educators (social, environmental, ethical, and cultural) arguing that challenging creativity through these limitations and dilemmas is necessary to provide learners with an education grounded in the twenty-first century context and demands (Craft, 2003).

Creativity is not only central to the social and economic development of society but to the progress in knowledge. It is important to nurture everyday creativity and develop more creative approaches in teaching, to empower our students to not only be innovators and creators of artifacts, but cocreators of knowledge. To best further creativity in education, we must not stick to a standardized pedagogy and assessment of creativity potential. Collard and Looney suggest that we need to be reflexive in our teaching in order to establish creative partnerships with students that allow better access to the creative process (Collard & Looney, 2014). Traditionally viewed as a fixed trait assessed via summative assessments, creativity is now perceived as a skill that can be nurtured in open learning environments that promote the structuring of knowledge.

Julio Ottino, who once studied art, is now the Dean of Robert R. McCormick School of Engineering and Applied Sciences. Gary Morson is a professor of Slavic languages and literature, but once intended to study physics. In their recent post in *The Chronicle of Higher Education*, both argue that educational practices that merge the humanities and sciences create "*whole-brain engineers and scientifically inspired humanists*." They suggest that these types of educational practices will foster more than just innovation, as these experiences will help individuals be more flexible and adaptable to global changes (Ottino & Morson, 2016). Ottino and Morson suggest that providing courses that bring different modes of thinking will cultivate the "*whole-brain*" experience.

Other scholars talk about interdisciplinary practices to unlock creativity (e.g., Including Art in STEM, Daniel, 2015). For example, Bradley found value in promoting transformative learning, where through creative experiences students self-discover and explore their 'inner emotional worlds' (Bradley, 2012, p. 130). Newell and Kleiman trust that when students are engaged in learning experiences where they feel safe to take risks, to collaborate and play, stimulate students' creativity (Newell & Kleiman, 2012). These creative experiences are discussed too from the authors of the Innovators DNA—experiences that promote associations, questioning, observation, networking, and discovery. Regardless of the need for flexibility of our educational systems that would facilitate these processes, there are not many engineering schools that provide learning environments for self-discovery, risk taking, exploration of 'inner emotional worlds' (Bradley, p. 130), or freedom in creative expressions. There is a vast amount of literature on developing and enhancing students' creative skills but these studies mainly examine design courses and capstone projects (Ottino & Morson, 2016). While assignments of a creative nature are more widespread than the formal literature currently indicates, difficulties lie in capturing publishable data under common standards of rigor. Recent conference papers offer numerous "creative" assignments for fundamental engineering courses, although many of these do not have well-documented outcomes (Bairaktarova & Eodice, 2017).

In their work on creativity and education, Csikszentmihalyi and Wolfe argue that creativity takes a long time and it happens within a system of cultural rules and with the support of experts (Csikszentmihalyi & Wolfe, 2014). In the case of creativity in education, the authors claim that creativity is a joint result of "*well-presented knowledge, interested students, and stimulating teachers*" (p. 181). Another tenet of the situated learning perspective, particularly applicable in engineering education (Johri & Olds, 2011), is that knowledge is constructed in practical activities of groups of people as they interact with each other and their material environments (Greeno, 2011). According to Greeno, this construction of knowledge is based on our experiences, guided by opportunities to explore, discover, construct, and create. While we learn and gain knowledge, we also find and shape ourselves. Through the discovery process, we also learn who we are and what we are good at. Nurturing discovery is helping individuals to find and express their true selves. The time we live in, the places we were born, and the people who raised us influence our values and ways of seeking the truth, or push us to be individually creative while seeking the new and undiscovered.

Many describe creativity as a mysterious process, a flow (Csikszentmihaly), a production of novel and appropriate responses to an open-ended task (Amabile, 2012). Personally, I have experienced periods of creativity when I reached a crossroad and needed to make sense of who I was and of the world around me—of what I have and what I have lost.

Reflection on Creative Experiences

I grew up in Bulgaria in the last decades of the Communist regime, in a political system that did not allow ideas and opinions to diverge from the communist doctrine. I feel beyond fortunate to have parents who showed me the truth behind the regime and provided me with many opportunities for self-discovery—singing in a choir, playing the accordion (still in my parents' house attic), making art with charcoals. . . all while also playing in the streets until dark (acting out and cooking with mud meatballs for the neighborhood grandmas), writing poems. It was through such unstructured play that I first fell in love with the shape and nature of objects. Each object I found in the attic—a pottery wheel, textbooks covered with dust, an unknown antique—signaled a new understanding of who my predecessors were and what they might have been interested in. Writing poetry was like playing with objects, virtually, making an imaginary connection in my mind; discovering that rhyming words and matching simple objects is something magical and complex. Later, at the Special High School of Mathematics, that love of objects transformed to the love of geometry, drawings, visual arts, and reasoning. My desire to study all poetry, visual art, and drawings prompted me to make my first important life choice. I had won many Bulgarian poetry competitions and was encouraged to join the oldest university in Bulgaria—the University of Sofia—to major in Bulgarian language and literature. But, I wanted to study Theater design and stage craft, however because of a school admission policy during that time (the communist regime) I was not allowed to apply to that school.

I speculated about what other programs I could enter. While my friends and family thought I would go into art and creative writing, I went to study mechanical engineering instead. After learning that mechanical engineering involved drawing and graphic design, I enrolled in my first year technical drawing class. I quickly learned that engineering drawings had nothing to do with my sketches, graphic design, and white and black charcoal artistic drawings. My education at the Technical University of Sofia not only provided me with a completely new set of skills, but more importantly, allowed me to explore and further my love for objects, geometry, and drawings thereby garnering a more holistic relationship with the world around me. The first sparkle of influence I had was when my professor of Solid Mechanics at the final exam asked some of us to open a box, pick up a part, and talk about the chosen object and what it does. The students who could not describe the mechanical objects and their function, although they may have performed well on answering the exam's theoretical questions, failed the exam. At that point on I was interested not only in the geometry and shape of humanly made objects but became eager to explore another world of engineering design where decisions are made about materials, functionality, and effectiveness of new products. I was also eager to know more how these tactile and tangible objects help us learn and practice engineering.

I will never regret my decision to study mechanical engineering. I view my engineering degree as a huge accomplishment. Becoming an engineer enabled me to interact with humanly made objects as a creator who brings them into existence in a variety of shapes and nature. My education in mechanical engineering gave me the knowledge and confidence to transform abstract ideas into tangible objects. My love for tinkering and playing with words slowly transformed into the love of tinkering with things. Working as a design engineer has allowed me to continually hone my visual reasoning and artistic skills while solving complex engineering problems.

Most of my experience has been in working in interdisciplinary teams, locally and globally, with people with different cultures, fields, backgrounds, and education. I have observed how my engineering colleagues use their "other" skills along with their technical skills. I thrived in these environments, thinking I had reached my full creative potential.

Then fortune required me to make another life choice when in the summer of 2009 my father passed away. My father was my inspiration in life. He studied psychology and was a pottery and clay technology teacher in the School of Ceramics in my home town for more than 30 years. Following my father's death, I learned about the Engineering Education PhD program at Purdue University and decided to pursue the degree in honor of my father. At the end of 2009, I left my design engineer position to join the School of Engineering Education. I considered this life changing move a creative one; becoming an engineering educator and researcher brought me even closer to my father—by sharing the same life experiences as he had, I now feel that I have him back. His influence reaches to so many of his students' lives. Furthering my education with graduate studies enables me to work toward understanding what motivates us to learn and how the material world helps in this learning process. Now, as an engineering educator and researcher, I feel privileged to have my father's life experience to reflect and build upon through all facets of my personal and academic research. Ranging from the exploration of humanly made objects to the education of my students, I strive to enhance their ability to explore and express their creativity, discover their own potential talent to ultimately bring their ideas to fruition. In a perfect symbiosis now, I play with both, and equally love tinkering with words and things, experiencing *flow*, while creating manuscripts or objects. I have found ways to do what inspires me most to do my best in life.

The Experience of Flow in Education

Mihaly Csikszentmihalyi is the Distinguished Professor of Psychology and Management at Claremont Graduate University, and the author of the highly cited book *Flow*. In his research Csikszentmihalyi investigates the question "*What makes a life worth living?*" He has found that pleasure and lasting satisfaction in activities bring us to a state of flow, a type of intrinsic motivation. When a person is completely involved in what they are doing, when the concentration is very high, when the person knows moment by moment what the next steps should be is considered, that person is in state of *flow* (Csíkszentmihályi, 2008). On the other hand, extrinsic motivation refers to doing something because it leads to a separable outcome (Ryan & Deci, 2000). Intrinsic motivation refers to doing something because it is inherently interesting or enjoyable, which, Ryan and Deci argue, results in high-quality learning and creativity.

Csikszentmihalyi represents the field of positive psychology, which investigates through scientific inquiry the strengths that enable individuals and communities to thrive. This field is founded on the belief that people want to lead meaningful and fulfilling lives, to cultivate what is best within them, and to enhance their experiences of love, work, and play. However, many of the tasks that we want our students to perform are not inherently interesting or enjoyable. Knowing how to support active and volitional (versus passive and controlling) forms of intrinsic motivation becomes an essential strategy for successful teaching (Ryan & Deci, 2000) and an opportunity for our students to freely express themselves in creative environments. According to Csikszentmihalyi, we are in flow experience when engaged in an activity that is appropriately challenging to our skill and confidence level, often resulting in task immersion and concentrated focus. He also states that flow can result in deep learning and high levels of personal and work satisfaction.

When investigating students' engagement based on the flow experience of 526 high school students, Shernoff, Csikszentmihalyi, Shneider, and Shernoff (2003) found that students were more engaged when the perceived challenge of the task and their own skills were high and in balance, the instruction was relevant, and the learning environment was under their control. Participants also reported high concentration, interest, and enjoyment (e.g., flow) when they work on individual and team projects as opposed to being lectured or taking exams.

If we examine the current undergraduate engineering curricula, the project-based approach is one that presents opportunities for students to work on projects where flow can be eventually achieved. However, in many engineering programs, students do not experience project-based learning until their last year of college in the form of capstone projects. In contrast, students in the liberal arts, for example, do not learn and master only their course material in the first 3 years of university—instead, from early on in the curriculum, they engage in their own unique ways of thinking to grasp opportunities that can contribute distinctive elements in their field. What can we learn in engineering education from the humanities to enable our students and ourselves to open the door to *flow?* Experiential learning is an educational approach where, through the exploration of real-life activities and challenges, students are involved in hands-on, collaborative, and reflective learning. Through learning from the process, students "take ownership" of the development of their new skills and knowledge. Learning environments rich with tasks that improve students' motivation and stimulate their creativity can also invite work that crosses disciplinary boundaries. In engineering education, experiential learning is mainly viewed as simply hands-on experiences. In reality, there is much more that we can learn from other disciplines when trying to design learning environments that improve students' learning, motivation, creativity, and appreciation of the subject matter.

A systematic review of the literature on successful educational practices for young people conducted in Scotland focused particularly on creativity in thought and practice with the aim of understanding how to effectively incorporate creativity into the curricula (Davies, Jindal-Snape, Collier, Digby, Hay, and Howe, 2013). The review has four main objectives:

1. What are the main aspects required to have effective creativity-stimulating experiences?
2. How will enhanced creativity training further the student?
3. What should the instructors do in order to 'teach creativity'?
4. How should those instructors be supported in their efforts?

The systematic review encompassed the literature from 2005 to 2011, and of 210 empirical studies only 58 were found that address the four questions earlier. Furthermore, only the first question had the largest breadth of data so the literature review shifted focus toward those answers. The authors reveal three underlying themes describing the key characteristics of creative learning requirements. The first is the physical environment: a space that is malleable and capable of change through shifting the furniture around and openness to one another's work. Learning environments also need access to resources that will stimulate creativity (design, arts, crafts, electronics, etc.). Outdoor classrooms, where the natural environment can foster creative skills in conjunction with indoor classroom teachings, were also identified in the review as stimulating a creative environment. The second important tool involved in nurturing creativity is the actual learning environment—considering how students learn new information and how they are expected to use it. For instance, looking at lecture-based environments versus project-based ones. Specifically, Montessori schools were shown to produce students that were more adept at original thinking. Students, who were given ownership and control in their learning, effectively did better in thinking creatively, possibly due to their familiarity with the notion from a younger age. Establishing confidence in students' learning ability is important so that they take control in how they learn in later years. With reduced structure in classrooms, students become more active in their own learning which also bolsters their understanding of new concepts. Moreover, with a more relaxed classroom environment, it is possible for tangential and emergent concepts to be explored, allowing for a greater appreciation for the interdisciplinary nature of learning.

Another enhancing tool, revealed by the literature review, is that of the relationship between learner and instructor. This relationship has been shown to greatly affect creative skills. The most important aspects of this relationship are flexibility, trust, and open-mindedness such that students are encouraged to carry out their ideas and trains of thought without an overload of direction. Additionally, dialog between the two is just as crucial in bolstering the creative mind; having accessible teachers allows for more in-depth exchanges in lieu of simply handing out instructions. Similar to the outdoor classroom, other nonschool environments have been shown to enhance creativity in students. For instance, museums, art galleries, science and technology centers, and the like have been shown as successful places where students can better explore their creative sides (Davies et al., 2013).

The New Renaissance Artificers

In this section of the chapter, I describe my creative ways of teaching and assigning creative projects in several fundamental courses in Aerospace and Mechanical engineering (Thermodynamics, Dynamics, Manufacturing processes). I apply project-based learning in all of my courses and, like most of my colleagues, find a place for creative assignments in design courses (Engineering Design Graphics, User-Experience Design). There is a vast amount of literature on developing and enhancing students' creative skills, but these studies mainly examine design courses and capstone projects. I set out to introduce a creative assignment for fundamental courses and study its impact on learning as well.

The Thermodynamics Class

Inspired by William Graziano, a professor of psychology, I adapted an activity that he demonstrated in his personality psychology course (Wesselmann, Kassner and Graziano, 2016). The work of Professor Graziano is innovative and cross disciplinary. To best explain how applicable his pedagogical approach is to other fields, I will share how Dr. Graziano's relevance-based activity made a positive impact on my Thermodynamics engineering students.

For 2 years, I taught Engineering Thermodynamics, a sophomore-level course, focused on the development and application of the First and Second Laws of Thermodynamics to solving problems from a variety of engineering fields. The course requires extensive use of differential calculus to interrelate thermodynamics functions. Finding Professor Graziano's relevance-based activity a great way to encourage my students' creativity and appreciation of what is considered a very abstract and difficult subject, I assigned a project at the end of the semester in which students were tasked to illustrate a thermodynamics concept covered in the course through a song, poem, short story, short movie, or comics. Students evaluated the entries and voted on their favorites in a tournament-style bracket system. As a result of this assignment, I have been serenaded by students with songs about the virtues of thermodynamic principles and have watched examples of thermodynamic events played out in film, all of which is arguably enough justification in conveying the success of this activity.

In both years, students loved the activity and noted in their course evaluations that the project let them think differently about thermodynamics and increased their appreciation of the engineering profession. Students also learned from this activity, particularly on the aspect of thermodynamics represented in their project. For example, students who wrote poems about entropy all had correct answers on the couple of problems related to entropy (Bairaktarova & Eodice, 2017). Students' creative entries were posted on the College of Engineering Social media. Many students, faculty, and alumni enjoyed watching how future engineers are showing passion about engineering and are engaged in original thinking.

Following are examples from students' creative works. For more examples, see the online Thermodynamics open education resource (http://ouopentextbooks.org/thermodynamics/). All students gave permission for their work to be displayed here and at the open educational Thermodynamics resource and each student retains the copyright to their own work (Figs. 1, 2 and 3).

Creative teaching efforts have enhanced a multitude of skills that students need to succeed in their academic lives and beyond. Literature reveals that when creative approaches are applied in the classroom, students have shown increases in motivation, focus, originality, social skills, and classroom engagement (Davies et al.,

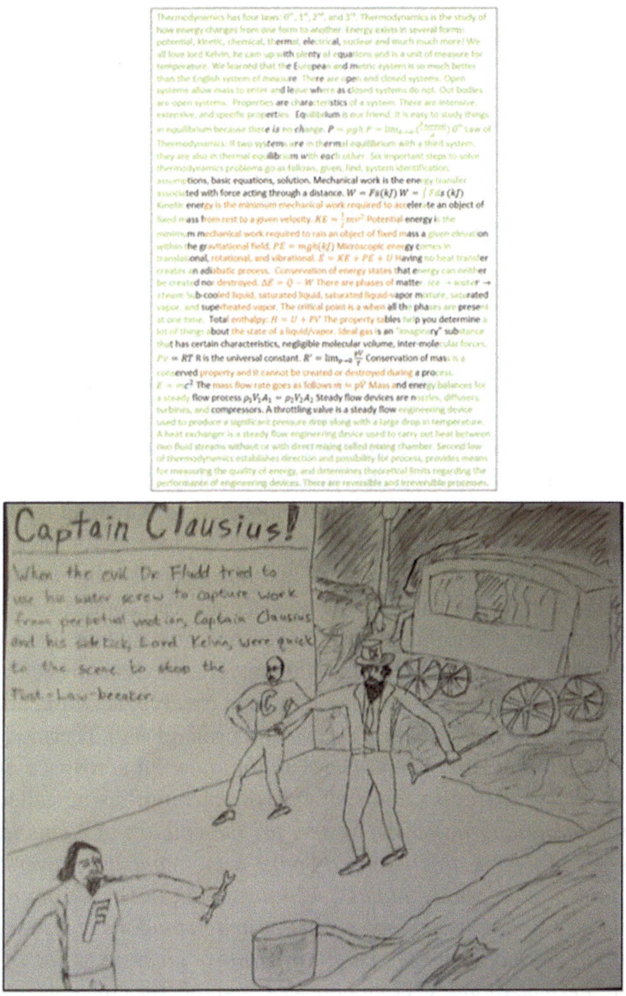

Fig. 1 Creative work, created by Jacky Bradshaw and Robert Chancellor

Thermodynamics Poems, Jokes, Riddles, and Nursery Rhymes

First and Second Law

Roses are red,
Violets are blue,
Energy is conserved,
And has quality too.

Entropy

Humpty Dumpty sat on wall,
Humpty Dumpty converted his potential energy into kinetic energy and had a great fall.
So scattered was he,
So high the irreversibility,
That, all the kings' horses, and all the kings' men,
Couldn't put Humpty together again.

Riddle #1

(answer on last page)
I am massive.
I absorb and release.
But still I remain constant.
What am I?

Fig. 2 Creative work, created by Christopher Sanders

2013). While increasing students' confidence, I have aimed to equip my students with the mind-set to be successful in many other aspects of their lives. Focusing on creativity in and outside of the classroom allows for a newfound comfort with the uncomfortable, thus enhancing the students' ability to use a more open and challenge-accepting frame of mind when confronted with a novel type of problem.

At the heart of any type of education is to prepare students for the real world. By allowing them to find their own ways of learning and applying advanced science principles, I am helping my students see how science can come alive outside of the textbook. By creating a learning environment where my students have the freedom and the time to make associations (explaining Thermodynamics in an animated Disney movie, created by Celeste Clay), question, observe, network, and discover, I hope I am not only teaching them the laws and principles of Thermodynamics but building on their Innovative DNA. Dan Carlton, now an aerospace engineer, has firsthand experience with my "atypical style of teaching," as he called it. As Dan shared: "*From designing a thermal system for the country of Namibia, which has budding energy problems, to creating a board game requiring players to correctly answer fundamental thermodynamics questions, I believe Dr. B's approach of more activities and less lecture has helped me better understand the complex and specialized subject of thermodynamics. Dr. Bairaktarova gives us a lot of freedom to explore what thermodynamics is and how we can apply the concepts in ways we can personally understand them. She encourages her students to continue applying their scientific passions through creative means beyond the classroom.*"

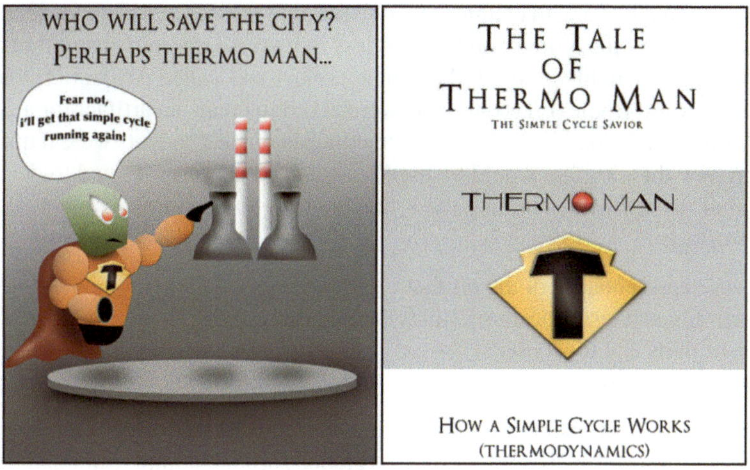

Fig. 3 Comics book, created by Jacob McAuliff

The Dynamics Class

Dynamics as described by many university catalogs is an undergraduate engineering fundamental course where students learn about the kinematics and kinetics of particles and rigid bodies for rectilinear, curvilinear, and angular motion. Students are also introduced to work and energy methods, conservations of impulse and momentum, and mechanical vibrations. Recently, I used a flipped classroom pedagogy to teach dynamics. This required students to watch my recorded lectures before coming to class to allow us to spend the majority of class time on problem solving in groups. The "flipped classroom" in my case could be used in the full meaning of the term "flipped." I have literally flipped the meaning and expectations of a traditional Dynamics course learning environment.

To encourage students' creativity in Dynamics, a course that applies mathematics to the study of forces, torques, and their effect on motion, and Kinematics, which studies the motion of objects without reference to its causes I have used an art performance lens to explain course topics. I have used everyday examples to explain concepts and increase students' confidence with understanding and applying the material and motivation for learning abstract concepts. I have twirled around like a skilled ice-skater (for sure I am not an ice-skater) in front of my students, wondering how many of them will know what I am trying to present. While some were thinking I am bridging together the arts and engineering for more engaged classroom, a few whisper softly that I am introducing the new topic of moment of inertia. While striving to find creative ways to engage my students in topics not always exciting, I also was trying to exemplify creative behavior and encourage students to find their own ways of associating course topics with their environment and life experiences, to observe and discover. In twirling around, I am also effectively breaking down teacher–student walls with counter self-preservation measures. The more accessible I am as a teacher, the more confident and less risk averse my students will become.

With these thoughts in mind, the course project was called Artifact Design (3D Motion of Rigid Bodies). The students were asked to create an artifact for a contest that illustrated a Dynamics concept covered in the course. Students were required to explain how their artifact would be used for educational purposes and to develop a manual of the artifact containing the following information to explain how the creation worked:

1. Visual representation of the artifact
2. Assembly and test procedure, safety instruction
3. List of parts and tools used
4. Note to the user

The creative project spanned 10 weeks. Students worked in groups of three, and the assignment was worth 20 % of the course grade. Through a mini-grant and support from the College of Engineering, I was able to bring the whole class to the local Science museum at the beginning of the semester. The mini-grant also supported the students' purchase of materials for their artifacts.

It was a very dynamic semester for the Dynamics class. From playing in the Science museum and getting inspired about ideas by learning about and creating the 3D rigid bodies, to immersing in a community experience—our Toy Fair, celebrating the power of creativity. The creative project not only served to help students clarify course concepts and encouraged them to think about the concepts in novel ways but it also promoted experiential learning through community engagement. Figure 4 below captures times of the semeters activities, such as students' play at the Science Museum, artifact testing in the classtoom work, and students presentation of their creations at the end of the semester Toy Fair. The same as in the Thermodynamics class, all students gave permission for their work and pictures to be displayed here and at the College of Engineering Social media and each student retains the copyright to their own work(Figs. 4 and 5).

I hope the experience my students had in the Dynamics class becomes a mind-set of inventiveness for the common good. The class invited professionals from the Science museum; teachers from the local elementary, middle, and high schools; teachers from the gifted and talented program; and faculty and students from the whole University. The College of Engineering documented the event to capture the creativity, confidence, and excitement of sophomores who were so proud of sharing their inventions and working on real project with a clause. Most of the projects were donated to Norman public schools, some were kept as presents from the students to their parents, I proudly keep a couple in my office, hoping to use these in harnessing creativity in the Dynamics classroom in years to come (Fig. 5).

From learning to draw a free-body diagram, applying concepts of kinematics and kinetics of particles and rigid bodies for rectilinear, curvilinear, and angular motion; generating ideas; working in teams; and understanding how things work in real life, students were engaged in the creative process and many times experienced flow as described by Csikszentmihalyi. They were experiencing the unified power of art,

Fig. 4 Play at the Science Museum and in the classroom and Toy Fair (Photography by Karen Kelly)

humanities, science, and engineering, while exploring their own talents and interests. When asked at the end of the semester what they liked about the creative project, 31 % of the class liked the project because of the freedom offered in creativity, followed by the 29 % who noted both the challenging hands-on and real-life application, and 11 % mentioned applying concepts learned (Table 1).

The Spatial Visualization Class: The Renaissance Engineer

The third example of applying a creative assignment is in seminar type introduction to engineering course. In addition to the lecture portion of an introduction to engineering course, 54 first-year engineering students were engaged in learning engineering sketching techniques to enhance their visual thinking. The major project for

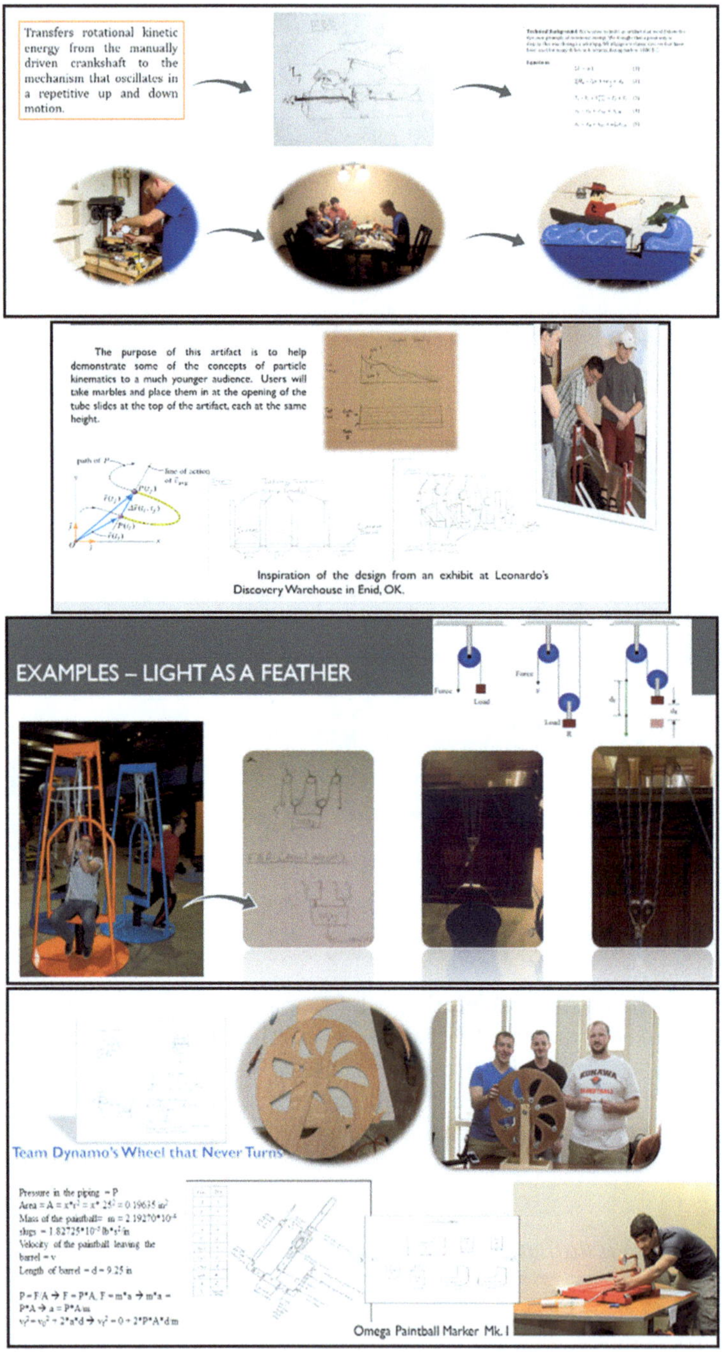

Fig. 5 Examples of students projects (development, artifact, and manual) (Photography by students and Karen Kelly)

Table 1 Students testimonies

Freedom offered in creativity
"I liked that we had the freedom to create any artifact that had dynamic principles."
"I like the innovativeness of the project. It is really good way to encourage creative thinking."
"I like the innovativeness of the project. It is really good way to encourage creative thinking."
"Dynamics can be a tough topic to comprehend, this project makes it easier to understand Dynamics in a fun way."
Challenging hands-on and real-life application
"This is the most hands-on project that incorporates real-life with school work."
"As a group we thoroughly enjoyed being able to apply the knowledge gained from our dynamics course to a real world application. It was an excellent way to see how dynamics works in a tangible way as opposed to just setting up equations and problem solving. We hope the users enjoy the artifact as much as we have."
Inspiration
When I was in elementary school, I visited the local science museum with a school group in the little town of Enid, OK. The museum had many exhibits such as animals, space shuttles, woodworking, pottery making, and dramatic role playing. However, the exhibit that I surprisingly was drawn to was a rolling ball slide where one was straight and the other had a wavy path. The amount of fun I had playing with the slide and the (then) surprising result made an impression to me. Even today, I am surprised that I remembered that exhibit and, while playing with our artifact, am still in wonder with the results. When we received this assignment, I wanted to make an homage to this exhibit, presenting the same concept except in our own way.

Fig. 6 Galileo's Geometric and Military Compass in Putnam Gallery (image taken from Wikipedia)

this section of the first-year cohort asked students to create an artifact manual. As an inspiration for their projects students were encouraged to look at Galileo's compass and the manual Galileo created (Fig. 6).

Through the semester, students were introduced to several contributions to engineering made by Galileo Galilei, including his first book *The Compasso* (*The Geometric and Military Compass*). They learned that in 1598 Galileo created a geometrical compass for military use. The geometric instrument enabled the construction and computation of the area of any regular polygon and circular sector. The students also learned that under Galileo's guidance, Marc Antonio Mazzoleni made the instrument and produced more than one hundred compasses. These compasses were sold along with an instruction manual that Galileo wrote in addition to offering a course of instruction in the use of the compasses.

Considering Galileo Galileo's "universal" intelligence, including drawing skills, I designed the course project to first excite students about freehand sketching, as research shows that sketching enhances spatial reasoning, but also aim to show students the interdisciplinary nature of engineering in their very first year in engineering school. In addition, I hoped this project will be a great way to engage the new students with the University community and work in a project aligned with the University's 125th anniversary, and more specifically the University Library Galileo's World exhibit. More explicitly, the project deliverable was an artifact manual, asking students to create a manual for an artifact illustrating an engineering concept. Similarly, to the Dynamics creative project, their manual needed to help middle or high school students to build the artifact and understand the engineering concept behind. Considering that experiential learning happens when through learning from the process, students "take ownership" of the development of their new skills and knowledge, not necessarily always engaged in hands-on project, to mention here, students were explicitly asked not to build physical artifact but to create an instructional manual. They were advised that their manual should be for an artifact that is easy to build using low-cost materials that are readily available at hardware and toy stores. As such they first needed to receive my approval for the proposed artifact and the involved parts. The students were not limited to specific engineering concepts but rather encouraged to use their imagination and creativity. The project was a group project and groups of four were arranged at the start of the semester. The objectives for this assignment were as follows:

1. Help students practice freehand sketching
2. Help student practice perspective drawing
3. Introduce students to engineering concepts
4. Encourage students to think about engineering concepts in novel ways
5. Increase students' appreciation for engineering
6. Increase students' appreciation for different than words representations in engineering
7. Help students engage in an experiential learning community

Through the semester students attended several times the rear books Science collection in the University library, had the Director of the History of Science collection as a guest-speaker in class, who presented the work of famous artists/scientist from the Renaissance time. Many meaningful conversations happened through the semester when students interacted and seek for advice and feedback on their projects from scholars in the library and the History of Science collection. The class evaluated the groups' manuals and voted on their favorites based on originality of the project, use of sketching, meeting the project required criteria, and quality of presentation. The evaluation rubric was based on the required elements of the project. Three winning groups presented their manuals to the end of the semester College of Engineering freshmen overall experience presentations. One project from our class is now included in the University Libraries Galileo Galilei's World exhibit.

Final Remarks

'Beauty is truth, truth beauty'

While I had not followed any framework or specific pedagogy aligned with any of the presented course projects earlier, I have strived to encourage creativity through students' exploration of their talents and to help them take ownership over their learning, learning environment, and more importantly over their creations. Be it a poem, short story, song or manual, real toy or simple machine—whatever the artifact may be—my students owned it. This is the case with the graduate courses I have taught as well. In the last chapter of this book, we will hear my graduate students' voices—the future engineering educators, who were enrolled in *Practicum in the Engineering classroom* graduate course. I have invited my students' reflections on creative teaching to be part of this collection.

So even if there is no tested or specific framework used in the presented above course projects, there are several plausible explanations why I trust these creative projects will be fruitful for learning in fundamental courses or any others. As Peter Brown and colleagues argued in their book *Make It Stick: The Science of Successful Learning*, a project could be very useful in that it adds a wild variation to the repertoire of students' retrieval practice (Brown, Roediger, and McDonald, 2014). Because doing a project like these described here requires an interdisciplinary mind and skill set, it is plausible to think that the knowledge students gain will be better retained. Further, these projects (and also many other similar efforts in education) present another paradigm of a flipped classroom. Flipped classroom often refers to a virtual space often designed for students to listen to or observe a lecture outside of class in a convinient for the student time and location, to latter be engaged with problem-solving activities in class. In an already tight engineering curriculum, there is usually no room for students to make presentations in fundamental courses. The thermodynamics and dynamics projects, for example, therefore present a good teaching model that could incorporate the crucial part of student presentations (learning by teaching).

With creative projects as the ones described in this collection, we could be reminded that creativity not only applies to the arts and humanities but to all disciplines, including engineering. While there is not always alignment between educational policies and implementation in practice (curricula), it is important that as educators we offer learning environments that enable students to develop and enhance creativity as a transferable skill and to create a space where creativity becomes "*an emergent property from interdisciplinary education*" (Craft, 2003).

To prepare the new renaissance artificers, we must learn from other disciplines, to design learning environments where students are free to observe, explore, collaborate, and discover their own potential. Many engineering subjects are abstract with complex concepts, overloading students with sophisticated knowledge. Students are more eager to learn when they see their classroom projects proving scientific principles rather than simply having to memorize principles found in text books. Students are more encouraged to explore and take risk when they see their teachers are artists themselves rather than as subject experts only—taking risks by implementing creative projects with outcomes not always straightforward to assess.

All authors in this collection, artists themselves, have included creativity in their educational goals, not being stopped by the fact that it is difficult to assess the outcome of creative projects. All chapters do challenge us to think about the necessity to enact a shift back to artistry in order to successfully encourage creative thinking and work in education. While students still have assignments and regular exams to complete, in our classes we seek to limit the lecture portion and emphasize group work and creative project involvement. We trust when students put their heart into a personal project—rather than their nose in a book—they learn more effectively. That is why for the projects described earlier in this chapter, the only parameter I enforced is that the project must explain a learned concept of the taught subject. How the students explain the concept is completely up to their creative whims. Creativity can be vastly different for each student. I appreciate each individual's efforts to turn the mountains of science and engineering into molehills of practical application and understanding.

From playing in conductorless orchestra, writing as knowing through science fiction, multiple messaging modes, optimizing a research paper in an Introductory Chemical Engineering Course, through encouraging creativity and identity in the construction of professional portfolios and provoking sometimes uneasy stories for critical reflection narratives and discussing ethical dilemmas in the engineering classroom, we are not only teaching students in enjoyable, innovative ways but also proving that learning creatively and learning science and engineering are not mutually exclusive. Students see that learning even complex subjects like thermodynamics and dynamics can be enjoyable. Just as John Keats proclaimed, "*Beauty is truth, truth beauty*," I see the beauty in the truth of art and engineering. My goal is to get students to express that beauty, whether that is reading about heat transfer in fluid flow, writing poems and singing songs about the principles of thermodynamics, or through creative ways of knowing, teaching others the beauty of engineering.

Acknowledgements The author wish to thank all students who participated with the creative projects and responded to surveys. I also wish to thank Kerry Magruder, for his involvement in the Spatial Visualization class and for stimulating many creative conversations on Renaissance engineering. Heartfelt thanks to the Director of Communications in the College of Engineering at The University of Oklahoma, Karen Kelly, for documenting and tirelessly promoting all the artifacts and events assoicated with students creative work in the classes described in this chapter. Finally, special thanks go to Vice Provost of Faculty Development Simin Pulat, Dean Tom Landers, and Associate Dean John Antonio in the College of Engineering at The University of Oklahoma for trusting my "atypical style of teaching"—your support contributed greatly to the success of the creative projects and students' openness to creative ways of knowing in engineering, overall.

References

Amabile, M. (2012). *Componential theory of creativity*. (Harvard Business School Working Paper, April, No. 12-096). Boston, MA: Harvard Business School.

Bairaktarova, D., & Eodice, M. (2017). Thermodynamics in high energy rhymes and rhythms. *Advances in Engineering Education*. (Forthcoming).

Bradley, F. (2012). Creativity: Does place matter? *London Review of Education, 10*(2), 145–157.

Brown, P., Roediger, H., III, & McDonald, M. (2014). *Make it stick: The science of successful learning*. Boston, MA: Harvard University Press.

Coate, K., & Boulos, A. (2012). Creativity in education: Challenging the assumptions. *London Review of Education, 10*(2), 129–132.

Collard, P., & Looney, J. (2014). Nurturing creativity in education. *European Journal of Education, 49*(3), 348–364.

Craft, A. (2003). The limits to creativity in education: Dilemmas for the educator. *British Journal of Educational Studies, 51*(2), 113–127.

Csíkszentmihályi, M. (2008). *Flow: The psychology of optimal experience*. New York, NY: Harper Perennial.

Csikszentmihalyi, M., & Wolfe, R. (2014). New conceptions and research approaches to creativity: Implications of a systems perspective for creativity in education. In *The systems model of creativity* (pp. 161–184). Netherlands: Springer.

Daniel, A. (2015). Full STEAM ahead. In *Prism, March–April, 2015*. American Society of Engineering Education.

Davies, D., Jindal-Snape, D., Collier, C., Digby, R., Hay, P., & Howe, A. (2013). Creative learning environments in education—A systematic literature review. *Thinking Skills and Creativity, 8*, 80–91.

Dyer, J., Gregersen, H., & Christensen, C. (2011). *The innovator's DNA: Mastering the five skills of disruptive innovators*. Boston, MA: Harvard Business Review Press.

Galluzzi, P. (Ed.). (1996). *Renaissance engineers*. Firenze, Italy: Giunti Editore.

Grainger, T., Barnes, J., & Scoffham, S. (2004). A creative cocktail: Creative teaching in initial teacher education. *Journal of Education for Teaching, 30*(3), 243–253.

Greeno, J. G. (2011). A situative perspective on cognition and learning in interaction. In T. Koschmann (Ed.), (Tran.), *Theories of learning and studies of instructional practice* (Vol. 1, pp. 41–71). Berlin: Springer.

Johansson, F. (2013). *Medici effect: What you can learn from elephants and epidemics*. Brighton, MA: Harvard Business Press.

Johri, A., & Olds, B. (2011). Situated engineering learning: Bridging engineering education research and the learning sciences. *Journal of Engineering Education, 100*(1), 151–185.

King, M. (2003). *The renaissance in Europe*. London, UK: Laurence King Publishing.

National Academy of Engineering. (2004). *The engineer of 2020: Visions of engineering in the new century*. Washington, DC: The National Academies Press.

Newell, A., & Kleiman, P. (2012). Doctors can dance. *London Review of Education, 10*(2), 133–144.
Ottino, J. M., & Morson, G. S. (2016). Building a bridge between engineering and the humanities. *Chronicle of Higher Education.*
Radcliffe, D. (2015). A tale of two steams. In *Prism*, December. American Society of Engineering Education.
Reznikoff, M., Domino, G., Bridges, C., & Honeymon, M. (1973). Creative abilities in identical and fraternal twins. *Behavior Genetics, 3*(4), 365–377.
Ryan, R., & Deci, E. (2000). Self-determination theory and the facilitation of intrinsic motivation, social development, and well-being. *American Psychologist, 55*(1), 68–78. doi:10.1037/0003-066X.55.1.68.
Shernoff, D. J., Csikszentmihalyi, M., Shneider, B., & Shernoff, E. S. (2003). Student engagement in high school classrooms from the perspective of flow theory. *School Psychology Quarterly, 18*(2), 158–176. doi:10.1521/scpq.18.2.158.21860.
Thermodynamics Open Educational Recourse Creative Commons (CC). Retrieved Sept. 25, 2016, from http://ouopentextbooks.org/thermodynamics/
Wesselmann, E. D., Kassner, M. P., & Graziano, W. G. (2016). Personality's top 40: An assignment to motivate students to apply personality concepts to their favorite songs. *Teaching of Psychology, 43*(2), 159–63.

The Engineers' Orchestra: A Conductorless Orchestra for Our Time

Diana Dabby

Introduction

Engineering students often view performance skills—leadership, teamwork, and communication—as "soft skills." Yet they are essential for advancing a professional career. To impart these skills, engineering educators recommend courses in the arts, humanities, and social sciences (AHS). Since a limited number of credit hours exist for nontechnical subjects in engineering curricula, educators focus on AHS topics most useful for engineers, many of whom will work in business and industry. The business community also values leadership, teamwork, and communication, often studying performing arts organizations as models worth emulating, such as the Orpheus Chamber Orchestra, a professional conductorless orchestra.

The raw material for these performing arts organizations can be found in colleges and universities worldwide; nearly all harbor musically talented students. Many of these talented musicians are also gifted engineers. Music-making has endowed these engineer-musicians with neurological benefits. Brain research has shown they are already primed for leadership, teamwork, and communication skills; thus they are excellent contenders for meaningful professional lives. It therefore makes sense for educators throughout the engineering community to nurture and encourage their engineer-musicians.

A project-based learning lab that builds upon the musical ability of student engineers can cultivate these skills necessary for professional and personal success. Such a lab exists at Olin College of Engineering. The Olin Conductorless Orchestra—an ensemble, minus conductor—features engineering students in collaborative, communicative, and leadership roles. It is the only conductorless orchestra composed of engineers—in the world. Yet as described herein, the rationale and blueprint for an orchestra where every member simultaneously leads and follows

D. Dabby (✉)
Franklin W. Olin College of Engineering, 1000 Olin Way, Needham, MA 02492, USA
e-mail: diana.dabby@olin.edu

D. Bairaktarova, M. Eodice (eds.), *Creative Ways of Knowing in Engineering*,
DOI 10.1007/978-3-319-49352-7_2

applies to diverse engineering schools large and small, East and West. Developed over the past 14 years, the OCO blueprint features a set of core values encouraging each player to be inquisitive, actively participatory, and ambitious. To emphasize these core values, OCO requires prospective members to rehearse with the ensemble as part of the auditioning process; at the same time, prospective members can ascertain whether OCO provides a good musical fit for them.

Other parts of the blueprint involve piece selection by the group, sight-reading sessions (where, for fun and challenge, students play music unknown to them), creative injections (brief humorous skits and activities that focus on some aspect of the orchestra or its repertoire), and board meetings to plan weekly rehearsal agendas—all geared to achieve peak concert performance. Professional string/wind/brass/percussion performers in the Boston area ("external guests") and a faculty "guide-on-the-side" provide constructive feedback.

Just as a self-governing orchestra such as the Berlin Philharmonic has myriad constituencies that influence its operation (Lehman, 1999), OCO has external constituencies, including the Olin Offices of the President; Provost and Dean of the Faculty; Board of Trustees; Admissions; Student Affairs; and Development, Family, and Alumni Relations.

What does a conductorless model offer student engineers preparing for roles in a twenty-first century world? In a word—leadership—and the responsibility that goes with it. That responsibility includes taking the initiative to diagnose problems, learning to effectively communicate possible solutions, and testing/implementing the discussed changes (Seifter & Economy, 2001). In so doing, the students build scaffolding for effective teamwork.

Background

A conductorless orchestra is first and foremost a large team collaboration by individuals working toward a high performance goal. By design, no overarching figure exists to sculpt an interpretation and pull the performance together. Rather, the members themselves bring the concert to fruition. Such an orchestra exerts singular demands on each individual during rehearsals and performances: the musicians must actively listen to their parts within the context of a larger whole and adjust accordingly, without being told by a conductor. A clear example entails all players watching one another to ensure everyone starts and ends together. More nuanced examples involve adjusting balance, dynamic levels, and tempo by listening intently, eyes on others, and cueing.

Professional conductorless orchestras have existed in the past: the Russian ensemble Persimfans (Khodyakov, 2008) and the American Symphonic Ensemble performed in the early twentieth century (Oja, 2000). Today a number of all-string conductorless orchestras exist, with the Orpheus Chamber Orchestra (1972–present) distinguishing itself from these by including winds, brass, and percussion. The Orpheus Process, i.e., how the orchestra evolved a play book enabling professional musicians to create a successful conductorless orchestra, has had significant crossover application to business and management practices (Seifter & Economy, 2001). Might a student conductorless orchestra composed entirely of engineers be viable? Can it help educate an engineer? The answer is yes to both questions.

The Basic Argument

Research has shown that for engineers to move forward in their careers, they require cognizance and experience with leadership, teamwork, and communication (Osburn & Stock, 2005; Seat, Parsons, & Poppen, 2001). But why design deliberate programs and/or course requirements to develop this skill set specifically in engineering students? Though millions of university students major in the arts and humanities, little concerted effort exists to offer programs furthering their performance skills, especially in leadership and teamwork. Why do engineers require special treatment? Because research has shown that engineers in general exhibit field-independent behavior characterized by a desire to work alone, a reluctance to engage in social activities, and difficulty in detecting nuance (Seat et al., 2001; Witkin & Goodenough, 1977).

Engineering constituencies in business, industry, and government have noticed this field-independent behavior. They want twenty-first century engineers, preferably with multiple intelligences that encompass, in addition to math and science, skills in language, leadership, and team building (Miller, 2008; National Academy of Engineering, 2005; National Science Foundation, 1996).

When industry, business, and government call for changes, the Academy listens. Engineering educators have worked with social scientists, psychologists, business leaders, and human resources personnel to create programs that address performance skills (e.g., at the University of Tennessee). Others have consulted with and hired performing artists to develop these same skills (e.g., Cooper Union).

The business community often examines performing arts ensembles, especially music ensembles, as models for optimal organizational performance. Such ensembles aspire to collective virtuosity—the ability of a group to reach peak performance (Marotto, Roos, & Victor, 2007). Students can also experience peak performance. Once experienced, peak performance becomes a highly desirable outcome that students want to experience again: it brings joy and "flow," as defined by Csikszentmihalyi (1996).

A student conductorless orchestra can reach collective virtuosity (i.e., peak performance by a group), as will be discussed with respect to the Olin Conductorless Orchestra. A necessary condition for collective virtuosity is transformational leadership, communication, and teamwork.

It is no accident that both the Orpheus Chamber Orchestra (a professional conductorless orchestra) and the Olin Conductorless Orchestra eventually evolved a distributed process of meaning-making leadership—a leadership style well suited for an array of teams in engineering, business, government, and academia. Such a process enables team members to find value in, and make sense of, their efforts[1] (Drath & Palus, 1994; Hackman, 2005; LaBarre, 1998; Seifter, 2001; Seifter & Economy, 2001).

Effective leadership requires excellent communication and teaming skills. Research has shown that music-making confers certain linguistic advantages to musicians. These include enhanced listening comprehension, hearing, and signal discrimination (Patel, 2007). A conductorless orchestra builds upon these attributes by setting a culture of when and how to listen, when to talk, and when to refrain.

Music may also be a key component for teamwork, as if it is wired into our human DNA from primeval times. Research by Dunbar (2012) suggests that music evolved in early human societies when societal groups became too large for social grooming. Music provided social glue. Similarly, a conductorless orchestra may offer a natural way to tap into human team building.

Neuroscience has found additional benefits of music-making. Male musicians, for example, have a relatively larger cerebellum and corpus callosum than do male nonmusicians (Hutchinson, Lee, Gabb, & Schlaug, 2003; Lee, Chen, & Schlaug, 2003). An increase in the size of the cerebellum suggests that its role in cognitive function and motor coordination is enhanced due to structural changes in neurons and their synapses (Holtmaat & Svoboda, 2009). Similarly, a larger corpus callosum indicates greater connectivity between left and right hemispheres of the brain (Lee et al., 2003). Therefore, a musician may already be 'hardwired' for skills transfer, showing an increased ability to make connections and transmit learning from one domain to another.

These neurological benefits indicate that the engineer-musicians in our midst are likely well suited for strong performance skills. A conductorless orchestra gives them a lab to experience and absorb shared leadership, cooperative teamwork, effective (and efficient) communication while doing something they love.

As an academic course at Olin, OCO works as a project-based learning (PBL) lab where students operate the classroom and the faculty instructor functions as a professional guide on the side ("faculty guide," for short). The members of the orchestra fill various roles such as 'piece leaders,' 'section leaders,' 'navigators,' and 'transcribers.' Elected piece leaders craft an initial interpretation of a work chosen by the orchestra. Section leaders, elected by their fellow players from each of the string, wind, brass, and percussion sections, help guide sectionals. Elected by the full orchestra as overall leaders, the navigators (usually two) work in conjunction with the faculty guide to ensure that rehearsals run smoothly and communication lines stay open within the group. For each new piece selected by the group, transcribers input the original score into Sibelius, a music notation program. The faculty guide then "reorchestrates" the original score to suit OCO's instrumentation for any given semester. Because Olin is a small school, its orchestra includes non-traditional instruments and unbalanced choirs of woodwinds, brass, and strings. The faculty guide also brings in external guests—musicians external to Olin who play

[1] In the words of Drath and Palus: "[Meaning-making leadership is] the process of making sense of what people are doing together so that people will understand and be committed."

professionally in the Boston area—so that student musicians hear additional perspectives on the music at hand. All OCO musicians work together with the navigators, piece leaders, section leaders, transcribers, faculty guide, and external guests to create a meaningful musical experience not only for the orchestra but also for its audiences.

In collaborating with each other, sometimes intensely and often with humor, the orchestra has risen to a higher level each year since its inception during fall 2002. The OCO rose to these higher levels due to an appreciation for leadership, skill, compromise, dialog, and respect for others. Furthermore, OCO students have consistently included some of Olin's finest engineers, mathematicians, and scientists.

A conductorless orchestra comprising engineering undergraduates can be successfully implemented at other engineering schools. The following pages show why and how such an orchestra makes sense for future engineers. They build an argument for a conductorless orchestra composed of twenty-first century engineer-musicians who will go on to collaborate, communicate, and demonstrate effective leadership throughout the world.

Motivation

We all have a tendency to view life through a particular lens which in turn reflects our habitual ways of thinking and being. Yet even a cursory reading of history and literature shows that in fact each day, any given situation offers many possibilities. A great leader and communicator has nurtured a "cultivated self" able to detect the nuance of possibility (Senge, Scharmer, Jaworski, & Flowers, 2005). How can engineering students reach their professional potential and a cultivated self?

In the 1955 Grinter Report, engineering educators advocated integrating the arts, humanities, and social sciences into undergraduate engineering curricula (Froyd, Wankat, & Smith, 2012).

> If the student is to be provided with a foundation upon which he may build a career of professional stature, his education must help him to seek his fullest development as an individual. [...]
>
> His facility with, and understanding of, ideas in the fields of humanities and social sciences not only provide an essential contribution to his professional engineering work, but also contribute to his success as a citizen and to the enrichment and meaning of his life as an individual. (Grinter, 1994)

In the years since 1955, the same appeals continue to be voiced. For example, in 1990 J. Ben O'Neal, then a professor at North Carolina State University, wrote a piece in the IEEE Communications Magazine where he observed:

> It is not uncommon to see engineers stranded in mid-career because of a narrowness of perspective and a lack of leadership qualities. Technical expertise is required, of course, and the universities do a reasonably good job of providing it. Most engineers are limited in their career not by a lack of technical knowledge, but by an inability to reason verbally, communicate their ideas to others, and furnish leadership. These skills seem to be better developed in the humanities than in engineering courses. (O'Neal, 1990)

Yet these appeals are not limited to the West. Calls for development of performance skills, specifically leadership, teamwork, and communication, are heard in many parts of the world. In a 1998 study comparing Turkish and North American engineering programs, Turkish educators noted the stronger emphasis on math and science at Turkish universities, but lamented the tendency to teach as many technical courses as possible while neglecting the arts, humanities, and social sciences. This predilection for the technical over the liberal arts has not only occurred in Turkey but elsewhere, presenting a "dilemma which a university in a developing country must resolve, if the aesthetic, psychological, sociological, and other cultural relations and consequences of scientific and technological development are to be taken into account" (Bilsel, Oral, & Pillai, 1998). Stating that the "importance of the humanities and social sciences in engineering education cannot be overemphasized," the authors quote O'Neal as to the reasons why:

> Humanities courses are also important for professional reasons. There is evidence that engineering and science courses do little to promote professional leadership. A [1986] study of leadership in the Bell System shows that, among people of comparable intelligence, those with an education in the humanities were judged to possess stronger leadership qualities than those educated as engineers. (O'Neal, 1990)

It is no secret, then, that the quality of performance skills in engineering graduates—including the ability to listen/write/speak effectively; resolve conflicts; work well in teams; respect diverse populations involving race, ethnicity, gender, and socioeconomic standing—concerns both industry and engineering educators (Seat et al., 2001). Thus, equipping and strengthening these traits in the undergraduate engineer becomes a necessity for all nations in an increasingly global environment.

Active listening, time management, exceeding expectations, flexibility, and respect for others are keys to leadership. All are developed in a conductorless orchestra. Without a conductor to cue them, musicians must listen and watch, dynamically adjusting to a constantly changing musical landscape. Outside of orchestra, these same engineer-musicians have to manage their time well enough to include practice and rehearsal sessions on top of an already rigorous academic schedule. Because the bar can always go higher in music, they know how to raise and exceed expectations. Yet each has to be adaptable and flexible enough to perform the group's repertoire even if the interpretation of a given piece, e.g., the tempo decided upon by the orchestra, goes contrary to what they personally feel. In this way, they demonstrate respect for others.

Field-Independent Engineers

To further provide context for the value and usefulness of a conductorless orchestra, we have to consider why engineers are often stereotyped as possessing excellent 'technical skills' while lacking 'people skills.' Most communication breakdowns involving engineers occur because the engineer focuses on the details of the subject at hand and then expects others to follow suit (Osburn & Stock, 2005).

The typical engineering graduate likes to solve problems, dislikes unpredictability, likes structure, responds to challenge/competition, and has been rewarded for individual work (Seat et al., 2001). But life is messy, dynamic, unpredictable—often due to human action or inaction.

Scholarly research has supported these stereotypes through examination of the usual cognitive style of engineers, a style known as field independent behavior (Seat et al., 2001). By contrast, field-dependent children and adults respond more to external social cues and social situations than those who are field independent. Field-independent individuals have an "impersonal orientation," show less interest in others, and prefer nonsocial environments. On the other hand, they work with greater autonomy, cognitive analysis, and structure (Witkin & Goodenough, 1977).

In light of the earlier research, it is not surprising that performance skills (leadership, communication, and teamwork) can pose problems for a number of engineering students. Given a predilection for finding the right answer, without ambiguity, they can become uncomfortable in situations characterized by unknown outcomes and unplanned responses, e.g., perceived disagreement or conflict (Seat et al., 2001).

A conductorless orchestra offers real-time experience with conflict resolution, decision-making, and brainstorming within an established culture—music—where give and take are part of the behavioral lexicon. In OCO, students resolve conflicts about interpretation by playing the various suggestions and then deciding on the basis of musical reasoning. 'Excellence' voting determines the outcome: each student votes for all suggestions he/she is comfortable with, rather than choosing just one. The OCO's co-navigators are also elected through 'excellence' voting.

Field-independent engineers can also be uncomfortable with movement, yet physical gesture provides ease and efficient communication. Individual or group motion while playing helps a conductorless orchestra synchronize. Not only does movement connect performers with one another, it also connects their music-making with an audience. Here is a performance clip of OCO performing Dvorak's Eighth Symphony at the 2014 Olin Exposition: http://bit.ly/OCO_Dvorak8. It demonstrates the communication (including gesture), teamwork, and distributed leadership enabling all to lock in for the exciting accelerando to the end.

Constituencies

Since the mid-1990s industry and governmental agencies have pointed out that engineering students 'get' math and science but many fall short on the social toolkit that will allow them to further their engineering talents. Performance skills enable engineers "to use their technical abilities as a part of a team, to understand conflict as a means for discussion instead of an angry confrontation, and to respect difference as a creative opportunity rather than an obstacle. To achieve success, today's engineers must be team players who thrive while working with a variety of people having differing social, educational, and technical skills" (Seat et al., 2001).

Then President of Exxon Research Clarence Eidt reinforced this view in a speech entitled "Institution-Wide Reform: Toward a Coherent Plan" given at an NSF-sponsored colloquium on revitalizing engineering education in 1996:

> "Appropriately prepared" Exxon employees have, of course, mastered the fundamentals of their chosen disciplines. But, they must have other qualities as well. We place a premium on employees who can communicate with those outside their disciplines, including non-scientists and non-engineers, who have the interpersonal skills for teamwork as well as leadership, who have the flexibility to grow and change as needs and demands change, and who will be able to serve the company in many different capacities over the years. We have to be aware of global interdependence in our world and an emphasis on teamwork over isolated individual activities. (NSF, 1996)

The 1996 NSF Report further supported Eidt's view by advocating student engagement in activities that encourage cooperation, teamwork, and discourse to increase both comprehension of concepts and appreciation for discipline. As a case in point, disastrous consequences can result from their absence, as occurred on January 28, 1986. Before the Challenger space shuttle exploded shortly after launch on a cold Florida day, Morton Thiokol engineers tried to warn their managers about possible failure of the O-rings that seal joints on the solid rocket boosters. They made their case technically but ultimately unpersuasively. The managers in turn failed to adjust their mind-set when presented with technical evidence. They failed to listen. If the engineers possessed more of a performing arts bent, they might have instinctively done whatever necessary to convey the direness of the situation (Osburn & Stock, 2005).

Performing artists specialize in communicating content, whether abstract (as in music) or concrete (e.g., a theatrical story). They engage all their senses, mind, and body to communicate directly with an audience. Reaching the public across footlights often defines the success of their art. In doing so, they become increasingly adept at communicating with different people, developing the ability to sense and therefore adjust to the expectations of those listening.

Arts-Based Learning

The business sector is always looking for ways to create and manage effective teams and to inspire and motivate individuals. By periodically reviewing new learning and development approaches, businesses help ensure their individual employees and organizational structures stay fresh (Manning, Verenikina, & Brown, 2005).

Just as the Academy (e.g., Cooper Union) has employed performing artists to develop leadership, teamwork, and communication skills in engineering students, the business sector has also looked to performing arts ensembles as viable models for organizational restructuring. This is quite a reversal of the usual business–arts relationship where art groups solicit the financial graces of corporations. But in today's world, that relationship has expanded (Seifter, 2001).

Why is arts-based learning effective? Why would Cooper Union and the business sector tap the performing arts to instill and motivate improved performance skills?

First, the process by which artists give feedback to one another, build teams, and rehearse carries over to business entities (Bartelme, 2005). Second, arts-based learning cultivates a "reflective intelligence":

> A valuable feature of aesthetic experience is the feeling of self-integration it induces. It is as if the harmony or unity of an artwork is replicated internally through a feeling that things have fallen into their proper places within a larger whole. This effect may be regarded as a kind of clarification, and it is often accompanied by the exhilaration that derives from the achievement of coherence among seemingly conflicting stimuli. The mind, as it were, is freed of clutter and can therefore act more decisively. (Smith, 1996)

Art fulfills the human desire for novelty and excitement by offering at its best, among the most pure and gratifying of human experiences. In addition to relieving tension and stimulating feelings of self-integration, aesthetic experience also develops perceptive and discriminatory capabilities. It stimulates imagination, which in turn fosters greater sensitivity to interpersonal relationships, e.g., by staying open, flexible, and cognizant of another's point of view. "Further consequences may be a disposition to entertain alternatives generally and to be more accommodating in one's outlook, that is, to transcend narrow-mindedness and stereotypical thinking. A cultivated aesthetic imagination ... will find application beyond aesthetic contexts" (Smith, 1996).

One such application can be found in the synergy between music performance and engineering practice. Both are lifelong arts. An engineer's world is always expanding as new discoveries and methodologies come online. The same applies to a musician: the better one gets, the better the imagination. The bar always moves higher. These disciplines necessitate collective diagnosis, often under the pressure of a deadline, e.g., a completion date for an engineering project and a concert date for a music ensemble. Deadline pressure often leads to the experience of "pulling something out of one's hat," i.e., finding a novel and fundamental solution to a pressing problem that might not occur without a deadline. Demonstration is critical to both disciplines. Engineers design, test, observe, and repeat the process till they get the best solution to the problem, given a set of constraints. Musicians test out different interpretations, revise, and refine in order to create a memorable performance. Finally, engineering and music demand practice—experiential practice. An engineer cannot be successful by book alone. He or she has to "practice engineering as a kind of performing art" (Felsing, 2007; Miller, 2008) to be an effective contributor. A musician has to step up to the plate and take action in order to make music.

Why Music Ensembles?

The business community has shown particular interest in the study of music ensembles, ranging from the string quartet as a self-managed team of four players (Gilboa and Tal-Shmotkin, 2012) to chamber and symphonic orchestras (LaBarre, 1998; Seifter & Economy, 2001). Businesses study these ensembles because, in the best cases, they successfully share leadership.

Writing in the *International Journal of Management Concepts and Philosophy*, Tovstiga, Odenthal, and Goerner argue that management can learn valuable lessons by studying musical performers and how they interact and communicate:

> In many respects, an ensemble is not unlike a modern business organization. Both exist in real time to generate value of some sort, whether for a concert audience or a group of customers. Both rely on complex organizational processes, involving interaction between individuals and the collective. Both rely on knowledge creation and its exchange, much of which occurs in the tacit, invisible realm. And, both can fail to generate value. Perhaps the only difference between the two is that "failure" in the case of an ensemble is immediately apparent; business failures may become obvious only after some time. (Tovstiga, Odenthal, & Goerner, 2005)

Leadership can often be flexible in music ensembles where players assume leadership roles as necessary. "Leadership may manifest itself in interchangeable roles (for example, variable and interchangeable lead roles are taken on by individual quartet members on an as-needed basis—often as dictated by the music score). Potential conflict, when it does arise, is typically worked out through playing rather than through talk" (Tovstiga et al., 2005).

Flexible, collaborative leadership also defines the Orpheus Chamber Orchestra, a professional conductorless orchestra which has given numerous workshops for the business community on successful shared leadership (Seifter & Economy, 2001).

Even a large conducted orchestra can have a form of shared leadership. Conductor Benjamin Zander has spoken at length on his orchestral role, consulting with business organizations worldwide. Here he presents his "I am a contribution" view of stewardship, a process he has employed with the Boston Philharmonic, an engaged orchestra with an avid audience base:

> Too much of the business world uses a narrow definition of success. I used it myself for a long time. I could not focus on what I had in front of me. I could think only about what else I ought to be doing, and whether that was enough. Then, one day, I had an epiphany. I realized that this is all a game we're playing. It's called "the Success Game"—or, I suppose, "the Success-Failure Game," because failure follows success everywhere. The Success Game runs in an endless win-lose cycle—which means that the people in it live with a sense of anxiety and fear.
>
> So I invented a new game, called "I Am a Contribution," or "the Contribution Game." It's easy: You wake up in the morning, convince yourself for a few minutes that you are a contribution, and you go out and contribute. Then you go to bed and do it again the next day. What I've discovered since I started the Contribution Game is that people have an endless amount of energy for it. Sure, goals can be energizing—when you win. But a vision is more powerful than a goal. A vision is enlivening, it's spirit-giving, it's the guiding force behind all great human endeavors. Vision is about shared energy, a sense of awe, a sense of possibility. That's what fuels the Contribution Game—and that's what's behind all great performances. (LaBarre, 1998)

Businesses find synergy with Zander's experiential views. Though he is the authority figure for the Boston Philharmonic and ultimately held accountable, he is also in the trenches with his musicians, doing whatever it takes to make sense and meaning out of what they accomplish as a community of practice.

Peak Performance: Collective Virtuosity

All of engineering's constituencies—industry, government, business, and academia—value peak performance. From a management perspective, working groups that achieve peak performance share certain attributes: their members experience something transformational, what Csikszentmihalyi describes as "flow," as being at "one" with what they are doing (Csikszentmihalyi, 1996). Peak performance often results in suspension of time, bonding the group together as they experience great joy and inspiration. Organizational literature has posited that "such experiences are gateways to increased creativity" (Marotto et al., 2007).

An orchestra provides an excellent laboratory for studying peak performance, which is by definition "action oriented." An orchestra's very existence depends on the ability of its musicians to bring masterworks to life and reach across the footlights to an audience. In a conducted orchestra, the conductor creates the character of a work by establishing tempo, monitoring balance, ensuring clean entrances and cutoffs, imagining a wide range of dynamic levels, timbres, and textures (Marotto et al., 2007).

But in a conductorless orchestra, the musicians themselves make these artistic decisions. For students who have played in conducted orchestras throughout high school, the conductorless orchestra offers a challenging experience and the potential for an optimum experience where the clock slows down, or even stops altogether, such that the players are "in the moment" (Csikszentmihalyi, 1996). Preparing for and presenting concerts offers excellent opportunities for students to experience the concentration and focus that enables "flow" to occur. Moreover, a flow experience in one domain can serve as a transfer mechanism for experiencing optimal performance in another domain (Brandsford & Schwartz, 1999).

A traditional conducted orchestra often relies on the charisma and leadership of the conductor to catalyze peak performance, whereas in a conductorless orchestra, everyone brings something to the table. Leadership and support emanate from each player. If someone does not contribute, then little benefit accrues for the individual or group.

Relationships among the members of the ensemble contribute to peak performance. Hearing a beautiful solo during rehearsal and/or performance galvanizes others to produce their best. This happens repeatedly in music ensembles. For example, the opening cello solo in a reorchestration of the popular YouTube hit Danzon No. 2 was cited by OCO students as 'setting the stage' for a transcendent performance:

> In between pieces before Danzon, I remember thinking about how much I had played already and wondering if I had enough left in me for Danzon. But as soon as Abe started in on his solo, all of those thoughts went away and I was engrossed in the music all the way through the big ending.

Group members are both soloists and active contributors in using their instruments, eyes, and ears to communicate with each other. Here a student describes his preparation:

> Before the first note, I looked around the ensemble, making sure to make eye contact with other players starting with me but still noticing the excitement in everyone's eyes. I'm still not sure whether the cause of excitement was nearing the end of the [full concert] or the piece itself. Maybe it was a combination of both. I vaguely remember feeling the beat and relaxing the tension in both my hands; it's almost like surrendering yourself to the music, letting the flow of the music itself take your conscience (or subconscience) away on a stream. I gave the cue, and we started playing.
>
> It was beautiful.

In communicating with one another, these student musicians form a community of practice. Their musical experience becomes collective. Collective virtuosity is defined as peak performance achieved by a group (Marotto et al., 2007), as evidenced by an OCO musician:

> I could feel everyone be a part of the music, and for the first time in a while, I was a part of a group of musicians who play *music*, not a set of musical notes. In fact, the performance of Danzon was the epitome of an ensemble playing music. Playing music that way transcends both the audience and the performers, and that experience is what I'd call a reward for an excellent performance.
>
> What started as a subset of orchestra and my solo carried on through the next ten minutes or so, where we were intensely focused and had the same exact goal and idea as to where we wanted to go and how we wanted to end this piece–*with a bang.*
>
> So, we did.

Collective virtuosity can emerge when the ensemble puts interpersonal differences into the background and its members focus on a larger purpose than themselves—the ensemble performance. In doing so, the group realizes that their sense of community and collective virtuosity can "co-exist with conflict, competition, and difference." In fact, how the group members interact with each other and engage with their work will ultimately enable or preempt collective virtuosity (Marotto et al., 2007). As another OCO student observed:

> There was a lot of leadership in Danzon, but even more support. For every solo and cue, there were a dozen other musicians listening, following, and providing backup. This was facilitated by strong communication and helped bring our performance to the next level.

Such engagement can happen across a stage. Interaction can even occur with the performers' backs to one another. In many OCO performances, the timpanist could be heard and seen leading from the rear of the concert space, giving 110 % of himself. As a result, the other players caught the energy coming from his timpani strokes and the precision of his beat. The timpanist was totally focused and at 'one' with the other players as he led them with his precise, clear beat.

Collective virtuosity can also be destroyed, as when a player enters at the wrong time. Then the spell or story being woven by the musicians is interrupted. Though the musicians catch on and the performance moves forward, the spell is broken (Marotto et al., 2007). So the player works harder for the next performance, as do the others: since they have each experienced flow at one time or another, they want to experience it again. Collective virtuosity is addictive—a joyful memory—"What made our performance most memorable to me was how we worked together to produce high quality music." (Student response May 19, 2013)

Teamwork

A conductorless orchestra provides a teaming experience. In fact, it resembles the Self-Managed Team (SMT) defined as a "group of employees that is responsible for a whole product or process, that does work that involves designing, evaluating and performing, takes possession of the results, and manages many of the things for which supervision or management are usually responsible" (Attaran & Nguyen, 2000).

SMTs under certain conditions can be very successful. They encourage shared and rotated leadership among team members. Each member accepts responsibility for team output and delivery goals, resulting in high commitment levels. Furthermore, "each team member must develop qualities of a leader such as risk taking and receptivity to new ideas and directions, in addition to technical expertise" (Gilboa & Tal-Shmotkin, 2012).

Many best practices for teamwork cross between the domains of business and the performing arts. To help a team reap the benefits of collaborative work and avoid pitfalls, at least one person must remain vigilant (Rouse & Rouse, 2004). In OCO those persons are the co-navigators and the faculty guide. Each brings experience from respective domains. For instance, the co-navigators understand the college's student culture better than the faculty member, while the faculty guide has professional experience.

For successful teamwork in a small ensemble such as a string quartet, the players must address three questions:

1. Designate a leader or pursue a democracy, i.e., decisions made by consensus?
2. What is the role of the second violinist? Typically the second violin parts are less demanding than the first violin parts yet the second violinist has to be as good as the first violinist to assure quality. This is why many quartets today switch the roles of the first and second violinist from one piece to the next.
3. If conflicts arise, do we choose confrontation or compromise? (Rouse & Rouse, 2004)

These earlier questions also apply to a conductorless orchestra. The OCO resolved the first by electing co-navigators who, among other duties, try to keep the group on task. The piece leaders and section leaders also help guide rehearsals and keep a good balance between talking and playing, with the emphasis on playing.

The OCO answered the second question by realizing that, at one time or another, each musician functions as a 'first' or 'second' violinist. As a result, the entire orchestra interchangeably fulfills these roles. The faculty guide reorchestrates pieces so that each player has solos and therefore experiences the importance of being supported—both musically and personally. When accompanying another musician who has a solo, OCO musicians listen to how the soloist is shaping the phrase, e.g., where s/he makes a crescendo, and then they adjust the dynamics of accompanying parts to 'lift up' the soloist, similar to how a male dancer supports and carries a ballerina through her role.

The typical OCO player (if there is such a thing) loves music, rises to a challenge, enjoys musical collaboration, makes suggestions, learns when to listen and when to speak up, likes solving musical problems (e.g., how to stay together through a tricky passage), learns how to give useful feedback and take initiative—all of which are among the necessary interpersonal skills of a self-managed team (Gilboa & Tal-Shmotkin, 2012).

Still, OCO has had its share of confrontations in the past. So to solve the third question, the musicians initiated a group audition to ensure not only that its musicians play well but also that they 'play well' together. Collaboration is central to excellence in any ensemble. Though the musical score provides expectations for the musicians, it does not spell out every detail; much room exists for interpretation. Listening to recordings of how other ensembles interpret a score can provide context (Rouse & Rouse, 2004). Both the score and recordings help resolve differences regarding interpretation, along with musical reasoning based on harmony, counterpoint, and orchestration.

But a larger question looms: Why might music provide an excellent way to achieve team building? In other words, *is there something in our human DNA that would make music a particularly apt conduit for teamwork?*

Among the many aspects of human behavior and cognition that can be counted as universals, music and dance are perhaps the most striking. Every culture boasts something that is instantly recognizable as music; every known culture dances. While there has been considerable interest in the psycho-biological underpinnings of musical ability and appreciation, and some interest in the timing of its origins, the one aspect that has proved unusually intractable has been its function (Dunbar, 2012).

Robin Dunbar, Director of the Institute of Cognitive and Evolutionary Anthropology at the University of Oxford, admits that though disentangling the evolutionary history of music poses difficulties, "careful consideration of the role of music in human social life suggests that its original function is most likely to have been associated with social bonding, once group sizes became too large to be adequately bonded using grooming alone" (Dunbar, 2012).

As a primate specialist, Dunbar also posits that music evolved from laughter, a vocal behavior unique to humans and chimpanzees. Just as laughter served as a grooming (bonding) tool, music also provided social glue and eventually, a natural bridge to language. The latter evolved as human group size increased toward levels where language inevitably became the mechanism for group bonding and cohesion (Dunbar, 2012).

Thus, current anthropological research suggests that music-making provides a natural conduit for human bonding. It follows that a conductorless orchestra offers a natural environment for collaboration, where students build, sustain, and contribute to a team.

Leadership

Common understandings of leadership center on two models: dominance and influence. In the dominance model, one person takes charge and others accept his or her leadership, at least initially, because the leader appears to have the best plan for proceeding forward. Winston Churchill in the early days of World War II embodied dominant leadership as he rallied his people to defend their nation and preserve liberty. In the influence model of leadership, the leader moves an organization forward by persuasion, by making the better argument. Abraham Lincoln coaxed the American Congress, sometimes member by member, to ratify the Emancipation Proclamation of 1863, freeing all slaves during the American Civil War. Leadership, whether by dominance or influence, usually focuses on the example of a powerful individual taking charge.

> This aspect of leadership is like the whitecaps on the sea—prominent and captivating, flashing in the sun. But to think about the sea solely in terms of the tops of waves is to miss the far vaster and more profound phenomenon out of which such waves arise—it is to focus attention on the tops and miss the sea beneath. And so leadership may be much more than the dramatic whitecaps of the individual leader, and may be more productively understood as the deep blue water we all swim in when we work together. (Drath & Palus, 1994)

An orchestra includes an array of knowledgeable musicians all focused on making good music together, yet all bringing varied personalities, ethnic, religious, and economic backgrounds into the mix. A leader of such a group might ask, according to the dominance and influence models of leadership:

a. How can I take charge of this talented and disparate group of people?
b. How can I influence them to work together harmoniously?
c. How can I make them accept my influence willingly, without having to resort to authoritarian methods (which won't work with them anyway)?
d. How can I make good things happen for this ensemble and accomplish our goals?
e. In short, how can I implement effective leadership? (Drath & Palus, 1994)

Yet another model for leadership exists that includes a meaning-making process—one that arises out of a group's realization and desire to make meaning, to make sense of what they want to achieve. As is well known, people resist committing to something they do not understand. One reason why the business world has studied artists for valuable lessons in leadership may be because artists have to reach out beyond the self to convey a vision. Thus, artists face an 'understanding' or 'making sense' barrier on a daily basis and ask, "How do I transform my private meaning into public meaning?" (Drath & Palus, 1994)

In a community of practice, however, a leader of the meaning-making process has to shift from the question "How can *I* make things happen?" to asking:

- "What do *we* need to do to make things happen?"
- "How can *I* figure out the best ways to participate in the process of *us* making it happen?" and
- "How can *I* help guide *our* activities toward the creation of significance?" (Drath & Palus, 1994)

J. Richard Hackman echoes (Drath & Palus, 1994) when he writes:

> A leader cannot make a team be great, but a leader can create conditions that increase the chances that moments of greatness will occur—and, moreover, can provide a little boost or nudge now and then to help members take the fullest possible advantage of those favorable conditions. (Hackman, 2005)

Leadership that not only focuses on sense- and meaning-making but also ensures that "the team's basic performance conditions *are* sound [thus helping] team members take the greatest possible advantage of their favorable circumstances" is the leadership model adopted by the Olin Conductorless Orchestra. This leadership is practiced by the co-navigators, piece leaders, and section leaders, assisted by each member of the orchestra and the faculty guide. Charged with enabling meaning- and sense-making to occur for the entire orchestra, the faculty guide and the co-navigators are ultimately held accountable for assuring the social and musical success of the orchestra. Thus, they might ask the following questions:

1. What is the nature of this group of people?
2. What is the most effective process of leadership for this group at this time? How might that process change as the group develops into a community of practice with a shared history that will unfold during this academic year?
3. How can we, as holders of some authority, participate productively in this process of leadership? (Drath & Palus, 1994)

In successful meaning-making, the leadership process becomes more distributed, resulting in influence as a beneficial outcome. Effective meaning-making leadership increases feelings of significance experienced by those in a community of practice. The question then for an authority figure is not how to get people to do what is necessary but rather how to participate in the structuring of the ensemble so that "people marginal to its practice are afforded the means to move toward the center of that practice. In other words, how can the contribution of each person in the community of practice be made increasingly important and increasingly appreciated for its importance?" (Drath & Palus, 1994)

In this "meaning-making" model, the leader(s) participate in the group, listen to its members as they discuss the task at hand, and understand the nature of their goal and its deliverables, i.e., why the goal matters, why the task at hand makes sense. Such participation helps lay the groundwork to create meaning for everyone's work. The leadership role becomes an empowering one, helping the group to make sense of the project and grasp its significance, or to paraphrase Hackman (2005), "creating the conditions for success." Understanding the group's goal as meaningful provides powerful motivation for everyone to work individually, and together. Their reward lies in the process of reaching their goal, and/or attaining the goal itself, thus bringing meaning to their professional or personal lives, and sometimes both (Drath & Palus, 1994).

A community of practice has a social character in that people work in tandem toward a goal. But "they are united by more than membership in a group or category." They are united by the "the power of shared activity to create shared knowledge and shared ways of knowing." Such a community, here a conductorless orchestra, focuses on the group's interpreting, anticipating, rehearsing, and planning together. Relative standing within the ensemble can shift depending on the topic at hand, according to an individual's time spent, expertise with that topic, and instrument (Drath & Palus, 1994).

Meaning-making can involve "influence" leadership when students take the advice/suggestions of other students in the group deemed more knowledgeable at a given time (e.g., piece leaders), as well as external guests and the faculty guide, whose combined professional experience can address remaining issues or concerns. Thus, together all participate in the sense- and meaning-making process.

Finally (Drath & Palus, 1994) ask: "So what about traditional leadership development? Should we abandon individual training in leadership? We think not."

> If, in the past, leaders have been trained to exercise leadership, they will now be trained *to participate* in leadership [emphasis added]. This is something like the difference between training an athlete in the individual skills of a sport and training that athlete in the team skills of the sport. Usually the individual skills are learned first. So it may be with leadership. Young supervisors and managers may need to learn the individual skills of leadership and later, as they approach higher levels of management, learn the community-oriented, meaning-making capacities, such as: (1) the capacity to understand oneself as both an individual and as a socially embedded being; (2) the capacity to understand systems in general as mutually related and interacting and continually changing; (3) the capacity to take the perspective of another; and (4) the capacity to engage in dialogue.

Sounds like a recipe for success in life where an engineer can navigate through uncharted waters, while fulfilling his or her potential at any given time. Through experimentation and feedback, the Olin Conductorless Orchestra evolved a distributed process of meaning-making leadership.

Communication

Clearly, our previous discussions regarding leadership and teamwork are predicated upon good communication, especially skill with speaking, writing, and listening. So one wonders: Does music-making develop language skills either directly or indirectly? The answer is yes. "Like language, music is a human universal in which perceptually discrete elements are organized into hierarchically structured sequences according to syntactic principles" (Patel, 2003). Furthermore, playing in musical ensembles enhances human hearing, enabling the ability to discriminate a desired signal from background noise. Studies also suggest that music can improve grammatical skills, aural comprehension, and the ability to differentiate between language that commands and language that questions (Enrenberg & Wolinsky, 2010).

Aniruddh Patel, author of *Music, Language, and the Brain*, studied the relationship between music and language from the standpoint of cognitive neuroscience. Patel's research has established that there are areas of the brain that process both music and language, i.e., music and language share deep and critical connections, both cognitively and neurologically (Patel, 2007).

As has been emphasized for at least 60 years, facility in expression—written and oral—is a professional necessity and an overall personal asset for engineering students (Grinter, 1994). Music-making activates many of the same brain areas as language and enables musicians to discern syntax. Syntax in language governs how sentences are built from words. A musical syntax guides how pitches and rhythms combine to form musical phrases in various styles, genres, and cultures. Whether processing musical or linguistic syntax, the brain appears to access the same neural circuits. According to Patel, "You can have overlap in the machinery that puts the pieces [the building blocks of language, i.e., words, or the building blocks of music, i.e., pitches and rhythms] together. They may be different pieces, but the machinery that puts them together is shared" (Enrenberg & Wolinsky, 2010).

Specific evidence provided by the imaging studies of Dr. Daniela Sammler of the Max Planck Institute confirms "a co-localization of the early detection of musical and linguistic syntactic errors within the bilateral superior temporal and perhaps in the left inferior frontal lobe, as proposed by previous fMRI [functional magnetic resonance imaging] and MEG [Magnetoencephalography] data. Overall, [our] present study adds (within-subject) [as opposed to between subjects] anatomic evidence to theories of shared syntactic processing in music and language" (Sammler et al., 2009). In other words, human detection of syntax errors in music and language occur in the same part of the brain, suggesting musicians may already be predisposed to effective language and syntax.

Additionally, by virtue of playing in a conductorless orchestra, musicians gain an appreciation for when and how to listen, when and how to talk, e.g., by acknowledging another musician's point of view. As is well known, both professional and student orchestras want to spend more time making music than talking about it.

Neurological Benefits

Why would it be in a nation's best interest to nurture its engineer-musicians? Why enable their talents with an Engineers' Orchestra, a conductorless orchestra to call their own?

Neuroscience research examining brain structure and activity has shown that music training changes the brain in lasting and positive ways (Enrenberg & Wolinsky, 2010). Research shows long-term motor skill activity results in structural changes to animal cerebellums. So one wonders what might happen to the cerebellums of practicing musicians.

A 2003 study found "a significant difference in absolute and relative cerebellar volume between male musicians and nonmusicians. Relative cerebellar volume correlates positively with intensity of musical training throughout life in the male musician group" (Hutchinson et al., 2003). Generally speaking, when a region of the brain increases in size due to some type of training (such as practicing an instrument), the increase in brain size due to the training is associated with stronger and/or a larger number of synapses, and perhaps even more neurons (Gage, 2002; Holtmaat & Svoboda, 2009; Reader & Laland, 2002). Conversely, deterioration of synapses due to aging or disease is associated with cognitive problems, memory loss, changes in mood, and other alterations in brain function (Morrison & Baxter, 2012; Terry et al., 1991).

Neuroscientists Lee et al. (2003) studied the relative size of the corpus callosum (CC)—the main band of interhemispheric axonal fibers in the human brain—in the brains of male musicians and nonmusicians. The corpus callosum functions as the key transfer and integrator of information between the right and left hemispheres of the brain. Lee, Chen, and Schlaug found that

> … male musicians had larger anterior CC than male non-musicians. This finding replicates the result of our previous study, and based upon anatomical and developmental research, suggests that male musicians might develop greater interhemispheric connectivity and increased hemispheric symmetry between motor areas and other frontal brain regions. The early commencement of intensive instrumental training during critical time periods of callosal development might trigger these differences. (Lee et al., 2003)

Daniel Levitin, author of *This is your Brain on Music*, summarizes earlier findings when he says that the human brain processes music throughout both hemispheres (Levitin, 2007). Thus, it is reasonable to assume that a musician's brain has developed more fully than if music training had not occurred. According to Levitin, as paraphrased by (Enrenberg & Wolinsky, 2010), "Playing an instrument, in particular, is an ensemble activity. It involves paying attention, proactive skills like thinking ahead, remembering, coordinating movement and interpreting constant feedback to the ears, fingers and, in some cases, lips." Clearly, music-making involves a complex array of processes. Neuroscientist Laurel Trainor, Director of the Auditory Development Lab at McMaster University in Hamilton, Canada, explains why: playing an instrument "engages basically most of your brain." Control processes such as memory, motor skills, attention, and executive function are all enhanced by music training (Enrenberg & Wolinsky, 2010). Furthermore, musicians typically develop highly desirable traits including discipline, steadfastness, dedication, and a certain comfort level with performing before audiences. In short, they are likely already primed for leadership, teamwork, and communication skills.

Skills Transfer

As noted in the previous section, neuroscience has already established that both the cerebellum and the corpus callosum are relatively larger in the brains of male musicians compared to nonmusicians. Therefore, a conductorless orchestra pulls together musically talented engineering students already predisposed to physiological transfer via enhanced interhemispheric connectivity of the corpus callosum and an increased number of neurons and synaptic contacts within the cerebellum. Such an orchestra allows these engineer-musicians, already predisposed to physiological transfer, to also experience transfer through the "preparation for future learning" perspective of Brandsford and Schwartz (1999). But first we have to define transfer. One definition is "the ability to directly apply one's previous learning to a new setting or problem (we call this the Direct Application theory of transfer)" (Brandsford & Schwartz, 1999).

Brandsford and Schwartz offer an alternative to the Direct Application theory. They call their perspective on transfer "preparation for future learning." It builds upon Direct Application theory by widening the scope to include learning that occurs when students solve problems via experimentation, research, critical commentary, and revision.

The "preparation for future learning" perspective values and assesses a student's ability to learn given the resources provided. "When organizations hire new employees they don't expect them to have learned everything they need for successful adaptation. They want people who can learn, and they expect them to make use of resources (e.g., texts, computer programs, colleagues) to facilitate this learning. The better prepared they are for future learning, the greater the transfer (in terms of speed and/or quality of new learning)" (Brandsford & Schwartz, 1999).

Preparation for future learning occurs when students first wrestle with a concept or design on their own, after which they become more receptive, curious, and appreciative of what an expert has to offer (Brandsford & Schwartz, 1999). A conductorless orchestra enables "preparation for future learning" because students first engage with the music themselves. Then professional musicians comment on their work. This process facilitates knowledge and skill transfer.

A "scaffolding of the learning experience" whereby students hear different points of view on a given topic promotes learning (Catterall, 2005). In addition, OCO supports this scaffolding on a weekly basis whereby students first learn from one another, suggesting and trying out different ideas for a given piece. Learning is then supplemented by other points of view resulting from different sets of musical experiences, e.g., audiences and professionals.

Finally, there is evidence of correlations between scientific success and musical hobbies that enable idea transfer. A 20-year study conducted between 1958 and 1978 by Bernice Eiduson examined 40 male scientists and their work habits, hobbies, time management skills, and attitudes toward the arts, humanities, and social sciences, as well as toward science. She interviewed these scientists (including four who went on to win Nobel prizes) four times during the 20-year period. Then in 1988, the 38 living subjects filled out a lengthy questionnaire regarding their "use of various forms of thinking (e.g., verbal, visual, kinesthetic), their avocations, forms and extent of physical exercise, and when they were most likely to have significant scientific insights (e.g., while working on a problem directly, while working on other problems, while relaxing, on waking)" (Root-Bernstein, Bernstein, & Garnier, 1995).

Eiduson then statistically analyzed the interview and questionnaire responses with respect to the scientific impact of each scientist to determine whether any correlations existed between "scientific success and avocations, preferred modes of thinking, use of time, energy, or related factors." The results indicated that

> significant correlations existed between scientific success and particular modes of thinking (especially visual ones), between success and various hobbies (especially artistic and musical ones), …, and between success and the efficient use of time to manage many competing vocational and avocational demands. We conclude that successful scientists have highly integrated networks of enterprise, whereas less successful colleagues tend to have fewer nonscientific activities that they do not integrate. They develop nonfunctional networks of enterprise in which activities compete against, rather than sustain, each other. (Root-Bernstein et al., 1995)

Before embarking on her research, Eiduson privately expressed "great skepticism that any connection might exist between hobbies, artistic proclivities, and scientific work." Thus, no positive interviewer bias existed with respect to the interview results. The survey revealed that "musical hobbies (e.g., musical composition, collecting records, singing, and playing an instrument) were generally associated with reported use of visual images while problem solving." The study also showed a "significant association between solving problems while working on different, related problems and being a high publication citation cluster scientist." One of these highly ranked scientists was asked if his hobbies affected his scientific work. He replied,

> [Suppose] someone is getting interested in musical problems. He may then apply what he finds there back to his scientific research. That's something which may affect very much the result. I think it's good. I think for a scientist who is working very hard, anything is good which brings from time to time another angle about general ideas into the picture. (#11, 1958). (Root-Bernstein et al., 1995)

Music, in general, and a conductorless orchestra, in particular, offer students such an angle, bringing various perspectives into their realms of thinking due to diagnosing and solving problems in rehearsals and practice sessions.

The Olin Conductorless Orchestra

What in the life of an Engineering School makes a conductorless orchestra possible? First, a universal call throughout industry for engineering graduates who can work well in teams, assume leadership, and communicate effectively. At Olin, these were founding precepts of the College. During the first week of meetings in fall 2000 when Olin's Founding Faculty of 12 sat together and started discussing possible avenues for the curriculum, these three performance skills were front and center. They were weighted heavily because the National Academy of Engineering and industry were insisting that engineers of the twenty-first century be multidimensional, i.e., able to not only work on multidisciplinary projects in large global teams but also poised to assume leadership and communicative roles. Much more would be expected of twenty-first century engineers than what had previously sufficed due to globalization and the exploding tech revolution.

Second, it is a well known though anecdotal truism across many nations that students gifted in math and science are often talented in music. Citing this, the Olin College Dean of Admission's first question to a newly hired Assistant Professor of Electrical Engineering and Music during the summer of 2000 was, "Okay, where's the Olin Orchestra?" This was not on the professor's to do list or contract. But the new Dean had concluded that an orchestra would be essential for a college that did not exist yet because of the often observed math–music connection. (Olin's inaugural class of students matriculated in the fall of 2002 and graduated in 2006.)

So what did a conductorless orchestra offer an engineering school? Why not a conducted orchestra? In the early days of Olin, the Leadership Team of the College encouraged the Founding Faculty to think imaginatively and try out bold ideas. They emphasized the importance for engineering students to function well in teams as conveyed by various constituencies. There were already several conducted orchestras at neighboring colleges and universities, any of which would be open to Olin students. Did it then make sense to start yet another conducted orchestra? Given the emphasis on teamwork, would a conductorless orchestra make more sense and have greater meaning for a new engineering college? The answers were yes and so OCO started in the fall of 2002 as part of a music composition/performance course with five students—on oboe, clarinet, bassoon, violin, and cello—a faculty clarinetist, and the faculty guide on piano. (The following year four trombones passed the audition and entered the orchestra, but there was still only one violin!)

Making Musical Decisions

Diversity in an orchestra includes instrument category, musical experience, gender, nationality, cultural attitudes, cultural values, and personality. This very diversity offers an orchestra the chance to develop a creative interpretation and understanding of the musical score.

Idea generation and idea integration are critical components for any effective team, and particularly a conductorless orchestra: they are predictors of successful ensemble performance (Boerner & Gebert, 2012). We find that OCO's diversity produces creative and motivational ideas that are often adopted by the group. These ideas are not only musical but also organizational, e.g., cueing is tackled by all and shows multiple points of view based on factors as mundane as where players are seated (sightlines), and more diverse factors such as instrument type and musical experience (some players have never cued before). Organizational ideas such as a group audition/orientation session and the implementation of co-navigators emanated from musicians representing different sections (strings, winds, brass, percussion)—each of whom had distinctive attitudes and values—yet all contributing toward the same goal: to design an excellent, well-functioning ensemble that enjoys making music together at an advanced level.

Many musical attributes need to be interpreted for the performance of a given work, including tempo, balance, dynamics, articulations, and expression marks. In a conducted orchestra, these are not typically discussed among all the players in an open forum. Yet OCO's full and sectional rehearsals center on these points. Technical issues germane to the different instrument families are resolved in sectional rehearsals. These include bowings (e.g., the bow arms of the string players have to move in the same direction for synchrony and coherence), articulations (e.g., the winds need to match attacks on notes so they come in together), breathing (e.g., the brass have to decide on a plan for staggered breathing during passages with long held chord tones). All of the above are necessary for attaining a cohesive ensemble sound. Yet the musicians have to remain flexible and motivated enough to develop and even change previously agreed upon musical and technical points.

A "Typical" Season

Each fall the orchestra passes through stages: rehearsing and performing with new members who have never played in a conductorless orchestra, working to prepare pieces for the first fall concert, listening to recordings of that concert to improve performance for the second fall concert (usually 1–2 weeks later), adding more pieces to the OCO repertoire in preparation for the Olin Exposition—a college-wide exhibition of student work. The best performance of the semester usually occurs at the Olin Fall Exposition.

Each spring the orchestra learns new repertoire and perfects the fall repertoire in preparation for 3 weeks of concerts for Olin's Candidate Weekends—a college-wide effort to welcome and evaluate selected high school seniors for Olin Tuition Scholarships. (Every admitted student to Olin receives a 2-year Olin scholarship.) Additional concerts occur off campus that feature the new repertoire and some of the fall repertoire, as selected by students. The best concerts of the year tend to be one of the Candidate Weekend concerts (standing room only for the past 9 years), an off-campus concert, and the final concert at the Spring Olin Exposition.

Figure 1 shows the OCO musicians prior to the first fall concert in October 2015. They quickly coalesced just 7 weeks into the term with Beethoven's *Prometheus Overture*, as can be heard in a videotape recorded by Joseph Hunter (Asst. VP for External Relations and Director of Communication) http://bit.ly/OCO_Prometheus. During spring'16, OCO attained another personal best with a complete performance of a major orchestral work—the final movement of Dvorak's *Ninth Symphony* in a concert for the Needham Lions Club http://bit.ly/OCO_Dvorak9th. (Though only audio exists for this performance, Joe Hunter created a "music video" using photos taken during AY15–16.)

Whenever the orchestra starts a new work, it passes through a rehearsal phase and then as the piece improves, a more motivated and creative phase. Finally, the orchestra reaches for, and sometimes attains, an inspirational phase, usually close to

Fig. 1 The Olin Conductorless Orchestra, fall 2015 (Photography: Michael Maloney)

a concert. Inspirational phases have occurred more often since 2013 in large part due to changes that encouraged bonding and shared leadership throughout the ensemble. The group built a cooperative environment defined by two dimensions: "warmth and support (mutual estimation) and a high level of cohesion (absence of conflicts)" (Boerner & Gebert, 2012).

Strategies and Blueprint

With each year, OCO strives to transform individual musical talents into collective creativity. To do so, its distributed leadership has to

- Recognize when change is necessary,
- Encourage new and fresh ways of looking at problems, and
- Initiate unconventional and innovative behavior (Boerner & Gebert, 2012).

When acknowledging that change is necessary, the orchestra restructures itself. To solve unproductive contention in the past, OCO changed its organizational structure, requiring a group audition for all members, adding informal and fun bonding activities, such as weekly OCO dinners, timely rehearsal breaks, and movie nights, as well as outlining new ways for repertoire selection. As an example of unconventional, innovative behavior, the co-navigators for 2012–2013 sent out rehearsal agendas and requests that always included something humorous in their sign-off. Figure 2 shows a photo that accompanied the following solicitation for feedback after a recent OCO concert:

> Let us know your ideas as soon as possible (while the performance is still kind of fresh in your mind)! Your musical rats, Philicia and David

Such unconventional emails resulted in a desirable outcome: students read and responded to them.

Fig. 2 Image accompanying an email sent by the co-navigators (Photography: © Ellen van Deelen. All rights are reserved.)

Gradually students and faculty alike learned that "the types of behavior in which people engage are infinite. But the range of useful behaviors, which make an effective contribution to team performance, is finite" (Belbin, 1993). Any orchestra, and especially a conductorless orchestra, has to set a productive, effective culture. Once set, it gets passed down from older to newer members. If it remains positive, then great benefits accrue. If it becomes negative, then problems bloom further and the orchestra has to reevaluate itself (King, 2006).

As many ensembles can attest, lethargic behavior and distracting personalities can derail an orchestra. Though these behaviors may occur from time to time, OCO's experience has been that their effects dissipate when others step forward into leadership positions during rehearsal—contributing energy, diagnosis, and suggestions on an as-needed basis. For instance, an often quiet student may speak up to rally a flagging rehearsal by suggesting a novel approach to the music at hand, sometimes with a quip. The orchestra has also devised activities through the years to keep rehearsals varied and exciting. Humorous skits and sight-reading enliven the group, giving players more spirit and focus for the work ahead. Fun and/or challenging activities called "creative injections" have become some of the most requested additions to a rehearsal. For example, students will mix up the different sections of the orchestra so that a French horn player sits next to a violinist or a flutist next to a cellist. Interleaving players from different instrumental families (a kind of "musical chairs") enables bonding with those from other sections of the orchestra. It also allows them to hear parts more clearly.

Interview data with respect to smaller musical groups shows that "establishment of a leader within a student ensemble is perhaps the most difficult, yet crucial part of building up an ensemble. Arguably, a leader can only fulfill his or her role effectively if the remaining members of a group oblige, allowing that person to take control" (King, 2006). The OCO avoids these issues because its leadership is 'spread out' across the orchestra. Over its 14 years, OCO developed a similar strategy to the (Drath & Palus, 1994) distributed leadership of sense- and meaning-making. Having flexible role behavior helped the group maintain an effective working environment (Belbin, 1993), where idea generation increased from a few players to virtually all. An appreciation and respect for setting a meaningful culture can stay with students the rest of their lives; they will bring these experiences to bear on future teams, group projects, and activities, whether in the workplace, among friends, while playing sports, or within their own families.

In sum, as the Olin Conductorless Orchestra evolved, so did its blueprint. In 2006, guidelines were established to give structure to the group. These included the positions of a navigator, piece leaders, and section leaders. As the orchestra grew and the competency of the musicians increased, the model evolved yet again, focusing on the following points:

1. An improved organizational structure now with co-navigators in addition to piece and section leaders and an OCO Board (all students invited to attend) responsible for drafting rehearsal agendas and mapping progress.
2. Core values that encourage each player to be inquisitive, actively participatory, and ambitious.

3. Clear expectations about commitment, attendance policy, shared goals, and productive behavior.
4. Individual auditions plus Group Audition and/or Orientation Session to ensure people not only have individual skill but also 'play well' with others.
5. Repertoire selected by members using 'Excellence' voting.
6. Bonding activities outside of rehearsal such as OCO dinner nights, movie nights, field trips, etc.
7. Fun (and surprise) activities during rehearsal such as "creative injections" and sight-reading.
8. Open communication lines for feedback on how things are going for each member in the orchestra. What's working well, what could be changed/more effective?

In a successful conductorless orchestra, each member not only has to talk the 'cooperative' talk but also walk the 'collaborative' walk. As Terry Bacon (2004) wrote in a piece for the *Journal of Business Strategy*: "You are how you behave." His is a powerful message for engineering students, many of whom will go into business-related fields. Yet the message goes far beyond business and applies to every facet of our lives. The old adage, "Actions speak louder than words" has a corollary in virtually every culture.

> Customers don't believe what you tell them. They believe what you do. Behavior is genuine. It is the purest form of the expression of your *intent*, your priorities, and your feelings. No matter what language you use to describe your business and products, no matter what promises you make or how sincerely you tell customers you want their business, the truth about you will always emerge in how you act. You are how you behave. (Bacon, 2004)

A Conductorless Orchestra as Project-Based Learning (PBL)

A conductorless orchestra offers a project-based learning (PBL) lab for performance skills—leadership, teamwork, and communication—much in the same way PBL has resulted in greater retention and understanding of technical subjects by engineering students (Froyd, 2011; Froyd et al., 2012).

PBL originated in the Medical School at Case Western Reserve University. It generally encompasses the following attributes:

- Posing the goal/problem before students have learned anything
- Providing "just in time" learning
- Empowering students with selection of learning goals, resources, assessment, etc.
- Enabling students to work cooperatively in small groups (with or without a tutor present in each group)
- Engaging students actively in the learning process, e.g., students teaching one another and giving feedback (Woods, 2006).

- Maintaining standards with a "Guide on the side," i.e., a teacher who monitors the process/progress and gives feedback

A conductorless orchestra encompasses all of the above. "Just in time learning" is provided by the piece leaders, faculty guide, and external guests. Students select their own repertoire before any music is learned. They provide feedback on the rehearsal process in real time and via email in preparation for the next rehearsal. The musicians work in one large group comprising all sections of the orchestra—winds, brass, strings, and percussion—and also in smaller units. Students teach one another since each brings a level of expertise to the music-making process.

The faculty guide has several roles similar to those responsibilities described by Drath and Palus (1994), specifically, doing whatever it takes to help the ensemble reach its goals and maintain standards. For smaller colleges such as Olin, the faculty guide also reorchestrates repertoire chosen by the orchestra. This is necessary because the students often select works originally written for large orchestras (80+ players). These pieces have to then be reorchestrated for much smaller ensembles (12–23 musicians) playing an array of instruments where the traditional balance among winds, brass, and strings is often askew. Knowing the capabilities and aspirations of each orchestra member, the faculty guide tailors the reorchestrations to create the best group sound possible with a nonstandard instrumentation. As an example, in his *Ninth Symphony* Dvorak wrote climactic passages that take the first violin section—typically 16–18 players—into the stratosphere. But in spring 2016 OCO only had six violins. Though the OCO violinists could technically play these parts, the results would have been thin and anticlimactic in comparison with the intended large violin section. So to achieve these climactic passages, the faculty guide scored the stratospheric lines for both glockenspiel and piano (right hand in octaves), doubled at the octave below by all six violins so that their overtones would support the glockenspiel (a high bell-like instrument) and piano. The results can be heard at the YouTube link http://bit.ly/OCO_Dvorak9th cited earlier.

As a PBL course, OCO undergoes a process analogous to a hands-on technical lab. As preparation for the hands-on lab (analog: an upcoming performance), students take a Pre-Lab test (individual practice to learn parts), assimilate Just-in-Time Learning (musical and interpretive analysis, e.g., through discussion), complete a Virtual Lab (orchestra rehearsals), and Hands-on Lab (concert performance).

As the final step in a PBL hands-on lab, students take a Post-Lab test. In OCO, the Post-Lab corresponds to follow-up rehearsals that address music or sections from the concert requiring more rehearsal as the group prepares for the next performance.

How does a conductorless orchestra as PBL move beyond the classroom? In the twenty-first century, teams will be increasingly tasked with providing solutions through multidisciplinary research rather than searching for answers within a single discipline. These complex problems will often require large teams. A conductorless orchestra enables student engineers to experience and contribute to a large group dynamic before entering the engineering work place, thus helping to jumpstart careers.

Possible Alternatives

If educators wish to create something close to a conductorless orchestra, they could begin by organizing chamber music groups comprising student engineers. Many of the same strategies and benefits provided by a conductorless orchestra also apply to piano trios (violin, cello, piano), string quartets (2 violins, viola, cello), piano quintets (2 violins, viola, cello, piano), and so on. Perhaps these chamber groups could eventually coalesce into a conductorless orchestra.

If no musical skills are available, other performing arts ensembles could be established, such as a theater ensemble. Again, many strategies and advantages already described with respect to a conductorless orchestra would apply to a theater group.

Though all of the above alternatives employ tried and true methods devised by educators and businesses for developing performance skills, these substitutes lack the cachet of a conductorless orchestra—a truly novel ensemble that catches the eyes and ears of others. At Olin, OCO has been affectionately referred to as "Olin's only varsity sport." It is unique in the world as the only conductorless orchestra composed of engineers, a statistic we hope to change. On the other hand, many of Olin's peer institutions have one or more choirs, *a cappella* groups, chamber music ensembles, and conducted orchestras. All of these ensembles provide meaningful and fun music-making experiences for students, yet none is unique. When OCO members go on job interviews, they are routinely asked by potential employers about their OCO experiences. A conductorless orchestra offers a singular experience to an engineering student and interviewers realize this.

Audiences also comprehend the distinctiveness of the enterprise. A conductorless orchestra can move and inspire constituencies such as alumni, parents, philanthropists, and audiences both internal and external to the university. For example, OCO performs every spring for each of three Olin Candidate Weekends sponsored by the Admissions Office. After one such concert, a mother of a scholarship candidate wrote in an email:

> I had expected to get a feel for the curriculum, students, physical plant, which I did, but I never imagined how much impact the performance of the Olin Conductorless Orchestra would have on my experience; it seemed to offer a beautiful symbolic representation of what the school is all about. I was moved to tears by the performance. I hope the members of the Orchestra know how truly amazing and transporting it is to hear music played collaboratively, without a conductor. From my perspective, the unique quality of the sound seemed to emerge from both the musicians' capacity to be intuitive about each other, modulating tonality and volume to achieve a balance, and a shared intense focus on the nuances of the pieces. I thought about how the qualities needed to achieve such a beautiful performance are what every good leader needs to cultivate.
>
> I tried to articulate something of what I felt to one of the orchestra members, but I think I was embarrassingly way too choked up to be clear—after all, I am a nervous mother with her first child going off to college! The musical performance was like no other that I have ever experienced. I hope that you will share my reflections with these wonderful musicians.

In comparison with coursework or workshops, a conductorless orchestra offers 'messier' problems. It exposes students to more unpredictable situations within a student-driven structure with scaffolding from faculty. Understandably, a self-managed team with 4–6 members is easier than a larger one, say with 22 members. Yet in today's world, the trend is toward larger teams, not smaller entities.

Fortunately, a conductorless orchestra model is transportable to other schools. Keeping in mind that OCO started off with just five students, a faculty clarinetist, and faculty guide at the piano, educators can successfully implement the model elsewhere. A blueprint already exists, as discussed in these pages, much of it directly transferable to other institutions.

The Engineers' Orchestra

The idea of "possibility," the notion of "what if?" lies at the heart of adaptability and flexibility, two characteristics demanded of persons young and old in this new century. Asking "what if?" implies an openness and curiosity about what might develop, what might ensue as a result of agency. And agency involves change. The power of "possibility" transfers to any domain. Possibility thinking encompasses problem solving and problem finding, including both convergent and divergent thinking. What constitutes "possibility"? Innovation, action, development, depth, risk, serious play, and posing questions all comprise necessary components of possibility thinking (Zander & Zander, 2000).

To date, a conductorless orchestra composed of engineers does not exist outside of Olin College. It represents a new idea. Yet performing arts ensembles, in general, and conductorless orchestras, in particular, are now studied by business leaders and academia as innovative labs for the development of performance skills—leadership, teamwork, and communication—and as models of organization for the business community (Bartelme, 2005; Gilboa & Tal-Shmotkin, 2012; Hackman, 2005; Lubans, 2006; Manning et al., 2005; Seifter, 2001; Seifter & Economy, 2001; Smith, 1996; Tovstiga et al., 2005).

Such performance skills cannot be viable without action on the part of the musicians. Music requires continued development as performers rehearse to bring works to the highest performance level possible. The bar is always moving upward. More broadly, bringing musical works to higher performance standards exemplifies lifelong learning, where lifelong learning is understood as a process that "encourages all adults to continue to seek new skills and knowledge." Lifelong learning remains a critical need for the world at large (Galal, 2008).

Furthermore, transfer as described by Brandsford and Schwartz (1999) occurs in a conductorless orchestra. Not only is depth required to continually revise and raise the performance bar for a musical work, but in the process, depth is also cultivated. "Depth transfer" can occur when professionals work with students, *after* students have first acted on their own to shape the interpretation of a given work. In doing so, students lay the groundwork for future learning from others who bring additional views and expertise to the table.

As any performer is aware, performance implies the possibility of failure. Audiences comprehend the success/failure dichotomy, one of the reasons why live concerts, as opposed to recordings, can have listeners on the edge of their seats. A conductorless orchestra involves more risk taking by student musicians than that required by a conducted orchestra. Many of the decision-making processes rest with them, such as cueing, staying together, and establishing tempo. During rehearsals, they answer one another's questions. Often these questions focus on trying out different interpretive ideas, hence the 'play' component inherent in possibility: playing with or trying out different ideas and combinations. For instance, OCO exemplifies the 'play component' when it rearranges its seating for better balance and group sound, or when it plays "musical chairs" by mixing up its seating for challenge and fun, as alluded to earlier.

A conductorless orchestra opens up the 'what if' question to educators. Such an orchestra can provide students with an opportunity to experience 'possibility,' take action, and bring something to fruition. It reinforces individual and group capacity for managing life's challenges, offering an environment where asking questions, assuming agency, and rising to the occasion are more the norm than the exception.

Playing in a conductorless orchestra is a lived experience that not only introduces student musicians to different styles of music but also to the varied backgrounds and perspectives of their peers. Students enhance their ability to work and communicate with others. Furthermore, the effective and "invested" leadership and teaming skills required in a conductorless orchestra can radiate outwards to the larger engineering student population and community as a whole.

Summary Remarks

Though engineering students, and even their professors, can view performance skills—leadership, teamwork, and communication—as "soft skills," they are in fact essential for advancing a professional career (Osburn & Stock, 2005; Seat et al., 2001).

Frank Barnes (1994 cited Bilsel et al., 1998), recipient of the 2004 Gordon Prize awarded by the National Academy of Engineering, summarized the reality:

> In terms of graduation, it is often the technology-based courses that help the students get the first job, the science course that helps them keep it five years out, and the social science and humanities courses that help them move up into top management.

An understanding and appreciation for leadership, teamwork, and communication prove necessary for engineers to advance in the profession beyond the entry level, i.e., beyond the fifth year of employment. *Performance skills allow engineering school graduates to reach their full potential as professionals, citizens, and individuals* (Grinter, 1994; NAE, 2005; NSF, 1996; O'Neal, 1990; Osburn & Stock, 2005; Seat et al., 2001).

Reaching one's full potential embodies much of human aspiration, yet in general, the student engineer is not born with these skills. In fact, engineers may be predisposed to the opposite side of the psychological spectrum. Research has shown that engineers often suit a cognitive style known as field independence, i.e., they prefer to solve problems alone rather than with others, they like unambiguous answers, and they favor nonsocial environments (Osburn & Stock, 2005; Seat et al., 2001; Witkin & Goodenough, 1977).

Engineering educators have pinpointed courses in the arts, humanities, and social sciences as necessary for giving engineers the tools to navigate problem solving, ambiguity, and social endeavors associated with multidisciplinary projects. Industry, nonprofits, accreditation, and government agencies continue to call for the cultivation of performance skills so that engineers can assume leadership, team, and communicative roles. Twenty-first century engineering projects will require these traits due to the global, multidisciplinary nature of the tech revolution (ABET, 1997; Grinter, 1994; Miller, 2008; NAE, 2005; NSF, 1996; O'Neal, 1990). Having these skills empowers engineering graduates to fulfill their promise as creators and contributors to society, enabling them to thrive professionally and personally (ABET, 1997; Erdil & Bilsel, 2005).

Consequently, many engineering schools have allocated a greater amount of the undergraduate engineering curriculum to the arts, humanities, and social sciences. Others have created programs that specifically address performance skills, including the use of performing artists to teach leadership, teamwork, and communication (Osburn & Stock, 2005; Seat et al., 2001).

The business community has also turned to performing arts ensembles to help restructure their organizations to improve the very same skill set. Study of performing arts ensembles has yielded valuable lessons for implementing distributed leadership, as well as models for successful teamwork and communication within an organization (Gilboa & Tal-Shmotkin, 2012; Hackman, 2005; Smith, 1996; Tovstiga et al., 2005).

In particular, the Orpheus Chamber Orchestra, a professional conductorless orchestra has evolved a collaborative process where leadership is spread across the entire organization, and communication skills are refined in order to enable efficient and productive rehearsals that lead to well-received performances (Hackman, 2005; Seifter, 2001; Seifter & Economy, 2001).

Significantly, a student conductorless orchestra at Olin College of Engineering has evolved a similar strategy for distributed leadership. The Olin Conductorless Orchestra functions as a conduit for experiencing peak performance (collective virtuosity), transformational leadership, teamwork, and effective communication. Overall, the OCO offers a sense- and meaning-making process of leadership that is shared by a community of practice (Drath & Palus, 1994). It also serves as a natural environment[2] for students to build, sustain, and contribute to a large team. Students

[2] Natural in the sense that current anthropological research suggests that music-making evolved as a medium for human bonding once a societal group became too large for grooming (Dunbar, 2012).

figure out when to say something and when to refrain. As a result, the orchestra spends more time making music than talking about it. Moreover, due to the absence of a conductor, students learn to actively listen. These performance skills transfer according to the "preparation for future learning" perspective of Brandsford and Schwartz (1999).

But why not just teach these competencies in other coursework? Certainly engineering educators, working with colleagues in business, the arts, humanities, and social sciences, can create programs to address performance skills. Yet if we widen engineering education scholarship to include the results of brain research, we find that recent work in neuroscience suggests the student musicians in our midst are already primed for developing effective leadership, teamwork, and communication. Neuroscientists have established that male musicians have a relatively larger cerebellum and corpus callosum than male nonmusicians. A larger cerebellum implies greater motor coordination and improved cognitive function, i.e., the ability to analyze and reason. A larger corpus callosum suggests more interaction and transfer between the brain's two hemispheres, since the corpus callosum connects both hemispheres (Hutchinson et al., 2003; Lee et al., 2003). Musicians also enjoy linguistic advantages over nonmusicians, according to neuroscientists Patel (2003, 2007), Sammler and colleagues (2009).

Clearly, engineer-musicians have become adept at time management, discipline, and the quest for excellence. Music is a demanding (and at times, unforgiving) art. Perhaps all the earlier traits account for why Olin faculty have consistently remarked that OCO includes some of the brightest and most engaging students at the College. These engineer-musicians are already wired for professional and social success. The raw material is present, ready to be further developed through bonding activities, individual practice, group rehearsals, and performances, all of which underpin a conductorless orchestra.

In short, our future leaders, team players, and communicators may very well be found among engineering students who also happen to play instruments. Thus, it is in the best interests of educators to nurture their engineer-musicians, encouraging them to take their talents along as they embark on 4 years of engineering study. It is a shame when students, facing pressure or disapproval from others, park their gifts in "long-term parking," only to return 4 years later to reclaim them. By then the keys are rusty.

Yet it does not have to be this way. In fact, great benefits accrue to students who take their musical talents with them as they proceed through engineering school and beyond. They can use these talents—their musical intelligence—to develop the essential performance skills needed to sustain, advance, and create meaning in a chosen career.

Musical intelligence constitutes one of the eight intelligences identified by Professor Howard Gardner of Harvard University, along with logical/mathematical, linguistic, spatial, intrapersonal, interpersonal, bodily/kinesthetic, and naturalist intelligences (Gardner, 1983, cited Miller, 2008). These intelligences reside in every human; they work together in practice to allow problem solving and invention within a given culture—much of what engineering is all about (Miller, 2008).

Citing the National Academy of Engineering's vision of engineering in the twenty-first century (NAE, 2005), Miller (2008) concludes that these eight intelligences will likely inform the engineer of 2020:

> ... to produce the engineer of 2020 we may need to broaden our focus beyond the traditional linguistic and logical-mathematical intelligence that is currently dominant in engineering education. Certainly these will continue to play the central role in engineering, but the other intelligences associated with artistic and social endeavors may need to be elevated significantly within our value system, in everything from the criteria for admission of students ... [to] the relative balance between technical and non-technical content in our academic programs.

The advent of a conductorless orchestra within an engineering program represents a new and ambitious endeavor that elevates additional intelligences to the same plane as the logical/mathematical. It brings together students from different ethnic, socioeconomic, and religious backgrounds. Such diversity strengthens the Academy and the profession. As is well known in the United States, diversity also brings challenges. Optimally, a nation would function as one team, with all working toward myriad goals that benefit citizens, en masse and in parts, taking into account the various needs of different regions and peoples. On the world stage, such an ideal team has been elusive because of the sheer size of the endeavor. But we already know that future projects in engineering will involve large multidisciplinary teams, solving messier and more complex problems.

The conductorless orchestra model offers a large team laboratory for students to experience and practice shared leadership, teamwork, and communication, while working towards definable goals—musical performances before audiences of peers, experts, and music aficionados. Such a lab would be a substantive complement to performance skills already being taught within the engineering curriculum. In fact, the two models (orchestral environment and classroom) could collaborate and grow stronger as a result of a two-pronged dedication to improving students' grasp of a valuable professional skill set.

When students of differing cultural and economic backgrounds have the opportunity to experience masterworks, i.e., great pieces of art, bonds form among them that transcend boundaries. Students experience both the creative process and the leadership process (one of making sense and meaning from the work at hand) all within the same ensemble, all of whose members have the same goals: performing classical masterpieces while experiencing the attendant joys and challenges.

As is recognized worldwide, classical music has its own culture—a transcendent culture that crosses borders and allows people to lead with, team with, and communicate with one another. Possessing a universal language and syntax, its lexicon has traveled far and wide. Thus, classical music provides a creative umbrella that allows diverse people to come together under the aegis of a shared culture and language—the language and culture of music. Moreover, when students have to adapt to new environments and cultures, their capability for skill transfer as "preparation for future learning" can deepen at a time when they have to reconsider or even let go of previous behaviors and beliefs. This kind of transfer allows more profound connections than the simple repetition of behavior in another environment (Brandsford & Schwartz, 1999).

In conclusion, establishing The Engineers' Orchestra as part of engineering curricula at home and abroad cultivates appreciation for—and experience with—leadership, teamwork, and communication. Our students will inherit and shape the future of all nations. In the historically recurring absence of effective governance, collaboration, and interaction across East and West, we as educators have to start somewhere. A conductorless orchestra provides access and opportunity for engineering students to lead, team, and express—three traits that will carry them far into this new century—for the benefit of all.

Acknowledgments Many thanks to Founding Librarians Lydia Carver and Dee Magnoni of the Olin College Library, Administrative Assistant Paul Coveney for his organizational and artistic help with the OCO concert programs, Asst. Vice President and Director of Communication Joe Hunter for his skill and creativity producing music videos of OCO, photographer Michael Maloney for documenting the Music Program at Olin with panache, artist Ellen van Deelen for allowing use of her imaginative photo in Fig. 2, Assistant to the Dean of Admission and Financial Aid Susan Johanson for generously giving her time since 2002, Founding Dean of Admission Charlie Nolan for his inspired and steadfast support of OCO, current Dean of Admission and Financial Aid Emily Roper-Doten for continuing momentum, President Rick Miller, Provost Vin Manno, and Boston University neuroscientist Judy Schotland for suggestions on the manuscript, and to past and present members of the Olin Conductorless Orchestra—all of whom made this orchestra a reality. On a personal note, deep appreciation to the OCO navigators of the past 4 years: Philicia Chow, David Pudlo, Jacob Kingery, Evan Dorsky, Shreya Rangarajan, Katie Hite, and Eric Westman.

References

Accreditation Board for Engineering and Technology. (1997). *Engineering criteria 2000* (2nd ed.). Baltimore, MD: ABET.

Attaran, M., & Nguyen, T. (2000). Creating the right structural fit for self-directed teams. *Team Performance Management, 6*, 5–33.

Bacon, T. (2004). You are how you behave: Customers can't be fooled. *Journal of Business Strategy, 25*(4), 35–40.

Bartelme, L. (2005). The view from the trenches: An interview with Harvey Seifter, & Tim Stockil. *Journal of Business Strategy, 26*(5), 7–13.

Belbin, R. M. (1993). *Team roles at work.* Oxford: Butterworth-Heinemann.

Bilsel, A., Oral, O., & Pillai, J. (1998). Turkish and North American engineering programs: A comparative study of curricular emphases on mathematics, basic sciences, humanities and social sciences. *IEEE Transactions on Education, 41*(4), 247–252.

Boerner, S., & Gebert, D. (2012). Fostering artistic ensemble performance: Exploring the role of transformational leadership. *Nonprofit Management and Leadership, 22*(3), 347–365.

Brandsford, J., & Schwartz, D. (1999). Rethinking transfer: A simple proposal with multiple implications. *Review of Research in Education, 24*, 61–100.

Catterall, J. (2005). Conversation and silence: Transfer of learning through the arts. *Journal for Learning through the Arts, 1*(1), 1–12.

Csikszentmihalyi, M. (1996). *Creativity: Flow and the psychology of discovery and invention.* New York: Harper Perennial.

Drath, W., & Palus, C. (1994). *Making common sense: Leadership as meaning-making in a community of practice*. Greensboro, NC: Center for Creative Leadership.

Dunbar, R. (2012). On the evolutionary function of song and dance. In N. Bannan (Ed.), *Music, language, and human evolution* (pp. 201–214). Oxford, UK: Oxford University Press.

Enrenberg, R., & Wolinsky, C. (2010). Music of the hemispheres: Playing instruments gives brains a boost. *Science News, 178*(4), 30–33.
Erdil, E., & Bilsel, A. (2005). Curriculum design to revitalise electrical engineering education at Eastern Mediterranean University. *International Journal of Electrical Engineering Education, 42*(3), 234–246.
Felsing, C. (2007). Introducing Dr. Richard K. Miller, meet our honorary board members. *American Suzuki Journal, 35*(4), 17–19.
Froyd, J. (2011). Problem-based learning and adaptive expertise. In *41st ASEE/IEEE Frontiers in Education Conference, 12–15 October 2011*. Rapid City, SD: IEEE. pp. S3B-1–S3B-5.
Froyd, J., Wankat, P., & Smith, K. (2012). Five major shifts in 100 years of engineering education. *Proceedings of the IEEE, 100*, 1344–1360.
Gage, F. (2002). Neurogenesis in the adult brain. *Journal of Neuroscience, 22*(3), 612–613.
Galal, A. (2008). New challenges facing the education sector in MENA. In *The road not traveled: education reform in the Middle East and North Africa* (pp. 83–114). Washington, DC: The World Bank.
Gardner, H. (1983). *Frames of mind: The theory of multiple intelligences*. New York: Basic.
Gilboa, A., & Tal-Shmotkin, M. (2012). String quartets as self-managed teams: An interdisciplinary perspective. *Psychology of Music, 40*(1), 19–41.
Grinter, L. (1994). Summary of the report on evaluation of engineering education [Reprinted from, *Engineering Education*, 46, 25–60)]. *Journal of Engineering Education, 93*(1), 74–94. Retrieved June 13, 2013, from http://www.asee.org/papers-and-publications/publications/The-Grinter-Report-PDF.pdf. (Original work published 1955)
Hackman, J. (2005). Rethinking team leadership or team leaders are not music directors. In D. M. Messick & R. M. Kramer (Eds.), *The psychology of leadership: New perspectives and research* (pp. 115–142). Mahwah, NJ: Lawrence Erlbaum.
Holtmaat, A., & Svoboda, K. (2009). Experience-dependent structural synaptic plasticity in the mammalian brain. *Nature Reviews Neuroscience, 10*, 647–658.
Hutchinson, S., Lee, L., Gabb, N., & Schlaug, G. (2003). Cerebellar volume of musicians. *Cerebral Cortex, 13*, 943–949.
Khodyakov, D. (2008). *Trust and control in counterpoint: a case study of conductorless orchestras*. Ph.D. thesis, Rutgers University, New Brunswick, NJ.
King, E. (2006). The roles of student musicians in quartet rehearsals. *Psychology of Music, 34*, 262–282.
LaBarre, P. (1998). *Leadership—Ben Zander* [online]. Fast Company. Retrieved May 16, 2013, from http://www.fastcompany.com/35825/leadership-ben-zander.
Lee, D., Chen, Y., & Schlaug, G. (2003). Corpus callosum: Musician and gender effects. *NeuroReport, 14*, 205–209.
Lehman, E. (1999). The berlin philharmonic orchestra. *Harmony, 9*, 11–25.
Levitin, D. (2007). *This is your brain on music*. New York: Penguin Group.
Lubans, J. (2006). The invisible leader: Lessons for leaders from the Orpheus Chamber Orchestra. *OD Practitioner, 38*(2), 5–9.
Manning, C., Verenikina, I., & Brown, I. (2005). Learning with the arts: What opportunities are there for work-related adult learning? *Journal of Vocational Education & Training, 62*(3), 209–224.
Marotto, M., Roos, J., & Victor, B. (2007). Collective virtuosity in organizations: A study of peak performance in an orchestra. *Journal of Management Studies, 44*(3), 388–412.
Miller, R. (2008). *Beyond study abroad: Preparing engineers for the new global economy*. Needham, Olin College of Engineering. Retrieved June 19, 2013, from https://olin.edu/about_olin/docs/pdf/study_abroad.pdf.
Morrison, J., & Baxter, M. (2012). The ageing cortical synapse: Hallmarks and implications for cognitive decline. *Nature Reviews. Neuroscience, 13*, 240–250.
National Academy of Engineering. (2005). *Educating the engineer of 2020: Adapting engineering education to the new century*. Washington, DC: National Academy Press.

National Science Foundation. (1996). Shaping the future: Strategies for revitalizing undergraduate education. In *Shaping the future: New Expectations for Undergraduate Education in Science, Mathematics, Engineering, and Technology Conference, 11–13 July 1996*, Washington, DC.

O'Neal, J. (1990). The humanities and their effect on engineering education. *IEEE Communications Magazine, 28*(12), 30–35.

Oja, C. (2000). *Making music modern: New York in the 1920's*. Oxford, UK: Oxford University Press.

Osburn, J., & Stock, R. (2005). Playing to the technical audience: Evaluating the impact of arts-based training for engineers. *Journal of Business Strategy, 26*(5), 33–39.

Patel, A. (2003). Language, music, syntax and the brain. *Nature Neuroscience, 6*(7), 674–681.

Patel, A. (2007). *Music, language, and the brain*. New York: Oxford University Press.

Reader, S., & Laland, K. (2002). Social intelligence, innovation, and enhanced brain size in primates. *Proceedings of the National Academy of Sciences of the USA, 99*(7), 4436–4441.

Root-Bernstein, R., Bernstein, M., & Garnier, H. (1995). Correlations between avocations, scientific style, work habits and professional impact of scientists. *Creativity Research Journal, 8*(2), 115–137.

Rouse, W., & Rouse, R. (2004). Teamwork in the performing arts. *Proceedings of the IEEE, 92*(4), 606–615.

Sammler, D., Koelsch, S., Ball, T., Brandt, A., Elger, C., Friederici, A., et al. (2009). Overlap of musical and linguistic syntax processing: Intracranial ERP evidence. *Annals of the NY Academy of Sciences, 1169*, 494–498.

Seat, E., Parsons, J., & Poppen, W. (2001). Enabling engineering performance skills: a program to teach communication, leadership and teamwork. *Journal of Engineering Education, 90*(1), 7–12.

Seifter, H. (2001). *The Conductor-less Orchestra* [online]. Leader to Leader, No. 21 Summer 2001. Retrieved June 1, 2013, from http://www.seifterassociates.com/uploads/4/0/5/9/4059165/drucker_foundation_-_the_conductorless_orchestra.pdf.

Seifter, H., & Economy, P. (2001). *Leadership ensemble: Lessons in collaborative management from the world-famous conductorless orchestra*. New York: Holt and Company.

Senge, P., Scharmer, C., Jaworski, J., & Flowers, B. (2005). *Presence: Exploring profound change in people, organizations and society*. London: Nicholas Brealey Publishing.

Smith, R. (1996). Leadership as aesthetic process. *Journal of Aesthetic Education, 30*(4), 39–52.

Terry, R., Masliah, E., Salmon, D., Butters, N., DeTeresa, R., Hill, R., et al. (1991). Physical basis of cognitive alterations in Alzheimer's disease: Synapse loss is the major correlate of cognitive impairment. *Annals of Neurology, 30*(4), 572–580.

Tovstiga, G., Odenthal, S., & Goerner, S. (2005). Sense making and learning in complex organizations: The string quartet revisited. *International Journal of Management Concepts and Philosophy, 1*, 215–231.

Witkin, H., & Goodenough, D. (1977). Field dependence and interpersonal behavior. *Psychological Behavior, 84*(4), 661–689.

Woods, D. (2006). *Preparing for PBL* [online]. McMaster University, Hamilton, ON. Retrieved June 15, 2013, from http://www.chemeng.mcmaster.ca/pbl/pblbook.pdf.

Zander, R., & Zander, B. (2000). *The art of possibility*. New York: Penguin.

Science Fiction as Platform for Problem-Based Learning and Teaching Writing as Design

Heather Marcelle Crickenberger

When we talk about "creative ways of knowing," it seems to me that we are talking about the generative nature of the learning process itself; we are talking about ways of interacting with something unfamiliar that are guided, not simply by a desire to understand what is already known about it, but by the desire to discover and create something new. As teachers who engage our students in creative ways of knowing, we open up the content of the class to outside influences that extend beyond our own knowledge and experience. Experimentation and play take the place of analysis and argument. Instead of asking our students to follow a procedure, where they wind up having done the very thing that was done before, we ask them to search out new ways of thinking about a subject, further expanding what is known.

The word *creative* can mean a lot of things. In the context of this collection, I see it as applying to both an approach to learning and a function of it. As a teacher of writing who has worked closely with engineering students for many years, I stand somewhat on the periphery of the discipline, but I'm close enough to see how creative it inherently is. Where I work, in Mechanical Engineering, students are literally creating things in their labs, and the writing they produce to document the process and analyze it is a kind of creative writing too, one that can be used to repeat or improve upon a procedure and better understand its significance to the field in general.

I've read thousands of technical reports and spent many hours talking to engineering faculty and students as part of the role I play as a technical writing advisor and English grader in our Mechanical Engineering Department. I've sat at length with engineering students to discuss improvements they can make in their writing, explaining our department's technical reporting guide. I worked on the committee that edited the guide and contributed to its content and I'm also currently working

H.M. Crickenberger (✉)
University of North Carolina at Charlotte, Charlotte, NC, USA
e-mail: hcricken@uncc.edu

D. Bairaktarova, M. Eodice (eds.), *Creative Ways of Knowing in Engineering*,
DOI 10.1007/978-3-319-49352-7_3

on a writing initiative to enhance engineering writing instruction college-wide. And while I have all this experience, I will never understand the equations I mark for formatting errors. I only vaguely grasp the theoretical principles of the thermal fluids and materials labs that I grade. While I can tell students they need to discuss their data more thoroughly, I do not actually understand their data. So the fact is, as a Writing in the Disciplines instructor, I can only meet my students half-way. However, when engineering students take classes in University Writing Programs, where I teach First-Year Writing to the general university population, they gain experience with design thinking and problem solving that will help them write more effectively in any discipline. What transfers from their general writing class to the writing tasks that face them in upper-level engineering courses is not content but a creative faculty. When students understand language as an implement of design and the task of writing as a problem to be solved at an early phase in their education, all writing they do becomes practice for the work that faces them in design-based fields like engineering.

This chapter chronicles how my approach to teaching writing shifted toward a design-based pedagogy as a response to interactions with students and professors in the visual arts and engineering. The chronicle demonstrates for others how to move into a design-based pedagogy, providing full assignments and examples of student writing; unpacking the writing studies concept of "Writing as Design" in relation to its Engineering counterpart; and proposing suggestions for structuring, implementing, and creating design-based writing assignments.

Summer 2002: The Artist Instructs

In the early 2000s, while beginning my web-based doctoral dissertation on the rhetoric of hypertext (University of South Carolina), a project through which I was exploring how hypermedia was changing the way we do scholarship, I enrolled in an Art Department summer class called *Advanced Drawing*. It was in this class that I first encountered the design-based assignments that would inspire my own teaching in the fields of English and Engineering by developing ways to help my students become more creative in the manner in which they approach the task of writing as a problem to be solved.

My motivation in taking the drawing class was primarily recreational. I always loved drawing and, as it was the summer after my qualifying exams when I was still at the early stage of writing my dissertation prospectus, I felt I had time to play. Because it was a summer class, our studio time was condensed to 5 weeks. We spent many hours each day in the studio. It was a large and open space, with tall bright windows and high ceilings. White drafting tables had been pushed together to make a labyrinth of creative surfaces. Most of the students were advanced Art majors and graduate students and were taking the class as a requirement. Our professor was known among them to be difficult. As an outsider, I found the atmosphere to be laid back compared to my literature and creative writing classes where everyone always seemed to be arguing. Most students worked in silence for the duration of the class, and apart from the little nudges and guidance he would give us as he moved around the room, our professor did not lecture.

Unlike many of the thesis- or question-driven assignments I had completed as a student of literature, assignments that required me to argue a point or interpret a text, this professor's assignments consisted of stringent and precise "Specifications and Requirements" (size limitations, required materials, structural rules, deadlines, techniques, etc.), which we were free to engage in whatever way we wished: so long as we met them all. Additionally, we had to be able to explain our creative process and thinking during our verbal critique sessions, and we had to reflect on our project in written form, responding to a number of the professor's questions. This piece of writing enabled us to explain how we met the Specifications and Requirements, in case they were subtle or difficult to recognize and required us to reflect on the work we had done.

I enjoyed this new mix of strictness and freedom. Instead of arguing a point or dissecting the writing of someone else, each assignment became a puzzle of sorts. For the first assignment, we had very simple instructions: *Do a drawing based on a grid using 30″×40″ paper, gesso, and oil sticks*. When we were done, we were expected to mat and frame our drawing with "professional quality" using only a razor blade and a ruler (a challenge in itself).

For the first few studios, we were expected to experiment with the media using different drawing techniques—cross-hatches, drips, finger smearing, etc. Gesso is a form of modeling paste that primes a drawing surface or can be used to build up texture, and oil sticks are basically large crayons made out of oil-based paint. When layered, they blend like oil paint and even the lightest mark is completely impossible to remove from a piece of paper. I had never worked with these materials before and had to spend a good deal of time experimenting with them before I felt like I had any control over them. I also spent a lot of time thinking about how I would base my drawing off of a grid. I wasn't sure if that was something the other students learned in Art 101 or if it was just literally supposed to be a drawing based on a grid, however I interpreted it. I ended up walking around Columbia quite a bit that summer, taking pictures of everything I could find that looked like it was based on a grid. I still have a folder on my computer called "In Search of Grids" with a few dozen images from these wanderings. I noticed grids in ventilation systems and saw their imprint in bicycle tires. They jumped at me off people's clothing and directed the rain toward the sewage system. After not long at all, I started seeing grids in everything—the pixels on the computer monitor—the chemical makeup of table salt. It seemed everything was based on a grid at some level.

In order to design a solution, I not only had to grapple with the language of the assignment, but also with the physical limitations of the required media and my lack of experience with the provided techniques; I had to work around the theoretical expectations of the assignment; and I had to articulate in writing how my project solved the problem posed by the professor, all the while attempting to create a finished piece that was visually pleasing, spontaneous seeming, and appropriate in form to my chosen subject matter.

After the allotted amount of time had passed, we presented our projects to the class for a mid-process "critique." We were allowed to revise or redo our projects before submitting them for a grade and, by that point, we were to have written out our answers to the professor's questions.

For the critique, I settled on using the grid to shape my light source. I did a portrait of a woman with a light shining through a grid-based pattern in a rattan chair I had in my apartment. My first sketches looked pretty good, but I wasn't aware the oil sticks had an effect that changed over time, and as the drawing set, I found I didn't like it and kept adding more and more layers of oil to the surface. By the time it was all done, the woman ended up looking as if she was in a confessional—and the manner in which the light speckled her face made it feel as if her shadows were many and small. I see the significance of this now, many years later, though at the time I was more concerned with training my hand to draw better portraits. Compared to some of the other art students, I felt like my drawing was lacking in realism. My proportions were off. When asked to identify the important aspects I was focusing on for the final drawing, I said:

> Mostly, I am trying to convey an emotion. It is a serious work, in that the expression on the face is serious and the piece is—overall—very dark. I want those who look at my drawing to feel the way this person feels—or to recognize that they have, at one time, felt that way…I am trying to pin down an emotion that can be caused by a grid. I am thinking mainly of the confessional—the grid that separates a person from their absolution—one that obstructs clear perception on either end, allowing the confessor and absolver to both hide…The light that shines through the grid onto the subject's face has been manipulated by this artificial structure. It is not whole, but scattered and fragmented… (Crickenberger, 2015c)

In this passage, I can see the influence of my literary studies in the way I interpreted what I was doing. The drawing was not very good. I still have it somewhere, buried in the back of a storage unit. When I remember it now, it's nothing like the work I currently do in liquid sumi ink and full-spectrum acrylic color. I don't see my style because I didn't have one yet. I was just learning how problems could be solved with these specific materials and techniques. I was simply a first-time wielder of giant crayons struggling with the ugliness of a grid. At the time, I had no idea that I'd be writing about the experience more than a decade later and confessing that it was perhaps one of the more influential assignments of my academic life.

Looking over my answers to the other questions, I am amused by the energy of the voice. I can see in the writing that I was excited to perform these experiments and to learn through problem solving—to experiment with a medium other than language. Figure 1 displays a few extractions from my first submitted assignment for the class.

Now, as a seasoned teacher, the assignment interests me from both perspectives: as an interesting model for structuring design-based assignments—one that combines rigid structure with playful experimentation in order to force creativity, but also as a student's response to such assignments. I can see learning taking place in the reflective writing. I can feel it happening again as I read the answers years later.

What stands out to me now is how generative the limitations had been. If the professor had just said, "Draw whatever you want," I'm not sure what I would have come up, but I wouldn't have come up with this. What I learned from doing this project and the other similar ones we completed throughout that summer is that I wanted to try a similar mix of freedom and rigidness in my own classes.

What new knowledge did you develop?

1. Through the manipulations of materials? First off, I learned that you get what you pay for…I messed up one of my best experiments by testing its dryness with my finger too soon. As I tried to fix the error, I learned that it can also be manipulated after being allowed to dry: the effect is totally different, of course, with much harder edges, but this could be useful….I am wondering if there is another medium that would enable me to build up texture that will stay. (I discovered later that yes there is: gel medium.)
2. Through thought? "The interesting thing about this experiment as how it makes you see the world around you. I became aware of the texture of everything. This is a good 'seeing' exercise….I also developed a new appreciation for AA batteries as artistic objects—who would have thought?

What does the problem ask you to do? Do a drawing based on a grid.

What are the important terms? Define them.

DRAWING: Pulling from—nearing (as in drawing to a close)—drawing a card or a straw—luck of the draw—drawing a response—to incite—to draw a crowd—to attract—*to render in a visual way on a 2-dimensional surface.*

BASED: founded, rooted, stationed, reliant upon, grounded in, anchored to

Fig. 1 "Project 1, Stage 1" Excerpts from my reflective Q&A paper for ARTS 330, 25 June 2002, Roy Drasites, University of South Carolina. The original response is formed from 11 of the professor's questions and is comprised of roughly 2000 words (Crickenberger, 2015c)

GRID: restrictions, a filter, a city grid, a way of sectioning off things so that they can be easily identified, something to plot points on (as in trigonometry), information that is organized in columns and corresponding rows, a repeated pattern that takes on the form of a square, order, structure, a system.

What prior knowledge is appropriate for the drawing and why? (from prior art classes or experience)…I was not familiar with the oil sticks, nor was I familiar with the idea of a grid-based drawing; however, I do feel I was able to complete the assignment with the use of common sense, as opposed to classical training….

What did you specifically learn from the problem? …I learned about the importance of light—that light is a thing, not simply an illusion—and that playing with light sources can give the artist different moods, different textures, and different situations. I also started to see that grids are all over the place—and that they are very often quite unattractive…which is why I used a grid made out of light. I wanted to look at what the shape could do if it were devoid of substance, if only a light were shining from behind it—a shadow of a grid

What are the strong and weak areas of your work? ...I know exactly when I should have stopped and I went ahead anyway...

Fig. 1 (continued)

Design-Driven Assignments

The following semester I started structuring my own First-Year Writing assignments as problems to solve. The assignments took the form of "Requirements and Specifications" (about 40 different requirements for a 2000-word piece of writing). I chose science fiction as a platform because most students were already familiar with these genres and because the genres themselves had been through so many iterations and rearticulations that they were rich with metaphors and recognizable conventions. It was also easy to familiarize students with the genres because examples of science fiction are so readily available. Students were prepared to bring their creative faculties to bear because the genre of science fiction is already widely associated with the creative processes of world building, speculation about the future, and storytelling.

The requirements and specifications were drawn from the conventions of various science fiction genres. Speaking generally, it is hard not to notice the way science fiction writers have guided empiric research with their creative imaginings. Mary Shelley warned us of the dangers of attempting to create life in *Frankenstein*. Mark Twain explored the political utility of scientific knowledge in a society of ignorance in *A Connecticut Yankee in King Arthur's Court*. Isaac Asimov famously pressed the logistical limits of the Laws of Robotics in *I Robot*, and current BBC series like *Humans* and *Black Mirror* have rendered visible some of the subtler dangers of networked living and convincingly humanoid machines. Science fiction gets its generative power from its use of narrative, perhaps the first and most natural way we make sense of the world around us. It is through narrative that writers are forced to grapple with complexity, to identify trends, patterns, and aberrations, and to speculate about the future and reflect on the past. It is narrative that enables us to share what we are thinking, but it is also narrative that shows us what we are thinking; and while all narratives are creative to some degree, science fiction has an added dimension—it moves us past the story of what is or was and into the question of what could be, opening up what we already know to the space of imagination and the possibilities.

As I finished up my doctoral studies over the ensuing years, I developed quite a few rule-based writing assignments around the genres of science fiction and fantasy, enlisting such atrocities as zombies, vampires, ghosts, and aliens into the realm of student options. These assignments were very popular among students, who engaged them as they would a game. I framed the assignments as "creative challenges," and while I could see students were learning something valuable by doing them, I hadn't yet found the language to explain what that was. Figure 2 displays a copy of my "Aliens" assignment.

At the time that I came up with this particular list of rules, I had already been using this format for my assignments for quite some time—many years in fact. Around the time that zombies started to suffer from over exposure, the *Twilight* craze had turned my vampire assignment into a gaudy display of sparkling teen cliché. The subject of aliens had yet to be exhausted. If you look at the rules presented in the assignment, you'll see many clichés of the alien genre. Primarily, alien stories deal with our anxieties concerning otherness and difference. They embody our fears about other ways of life—they place humans at a different place within the food chain or deracinate them from the known confines and comforts of Earth. Primarily, these stories deal with the concept of alienation, familiarity, and home. One sees these themes at work in classics like *E.T.* and *Star Trek*, but they are at work in any story that represents travel through unknown terrain: *The Odyssey, Alice in Wonderland, Dante's Inferno*—even *The Wizard of Oz*—are all variations on the alien narrative, in the way that they envision no place to be quite like home and in how they embody our encounters with the unfamiliar. On an even darker side, we can see how alien stories give form to our deepest fears about superior intelligences and physical capabilities; they deal with colonialism, predatory behaviors, and the sublime terrors of the unknown.

While designing this assignment, I chose to ban the word *alien*. I often ban the word that serves as the title of my genre assignments in order to encourage students to explore subtler, less overdone manifestations of the genre and to get them thinking metaphorically. I ask them to write in present tense because to tell a story that way is counter to our nature, and I want them to inhabit a tense that feels uncomfort-

Pop Culture as Social Critique: "Aliens"

Requirements and Specifications:

Write something that follows the rules below in less than 2000 words. Note all rules in your copies of the paper using the parenthetical codes listed below. Check off this rule sheet and attach to the front of my copy, along with a cover letter reflecting on your process and the product of your work, posing at least three questions you would like feedback on during workshop discussion. Bring 23 copies total to the start of class on the due date. Your paper must be formatted according to the syllabus. Papers that do not follow all rules cannot receive a passing grade. Papers that follow all the rules cannot receive a grade below a C. Paper 2 should be completely new work—no sequels or continuations of Paper 1 please!

1. ____ Do not use the word (or any derivative of the word) "alien" anywhere in the paper.
2. ____ Primarily present tense narration
3. ____ Difficulties of communication must be described (DIF)
4. ____ Someone/thing in the story makes a gesture that only one other character in the story understands (GEST)
5. ____ Fear manifests as violence (FMV)
6. ____ A common thing takes on great significance (SIG)
7. ____ The concept of loneliness must be addressed (LONE)
8. ____ You must have a character that dwells in two worlds (DWELL)
9. ____ A detailed description of a home from 2 different perspectives (HOME 1, HOME 2)

Fig. 2 "Pop Culture as Social Critique: Aliens" (Crickenberger, 2015a)

10. ____ Something must glow in the dark (GLOW)
11. ____ A character fumbles with technology he/she/it does not understand (TECH)
12. ____ Something hatches (HATCH)
13. ____ An aerial perspective (AREAL)
14. ____ A means of transport must be described (TRANS)
15. ____ Something slimy (SLIME)
16. ____ Something hairy (HAIR)
17. ____ Written symbols that are not understood by everyone in the story (SYMBOL)
18. ____ A higher mind/greater or unusual intelligence of some kind must be represented (INTELLIGENCE)
19. ____ Many eyes (EYES)
20. ____ A description of the night sky (SKY)
21. ____ An object is used for something other than its intended purpose (OBJECT)
22. ____ A cover-up (COVER)
23. ____ Five references to lighting or weather (L/W 1-5)
24. ____ A detailed description of someone's hands (HANDS)
25. ____ Five references to smell (SMELL 1-5)
26. ____ Five references to sound (SOUND 1-5)
27. ____ Five references to a texture (TEXT 1-5)

Fig. 2 (continued)

able so they can learn to be comfortable in it. (Raising student awareness about tense has a noticeable payoff in engineering, I learned much later, as it is the tense of theory and analysis, a common place of error in technical reports.) I thought about all the alien movies I had seen when designing the assignment—and I thought about all the books I read about alienation. When I look at the assignment now, I see how it can be used to tell any story about "a fish out of water" or the "fear of the unknown."

These early experiments proved useful when, in 2007, I moved to the University of North Carolina at Charlotte where I continued developing these assignments in the context of our First-Year Writing Program and the semester-long inquiry project that all lecturers were required to include as part of their courses. I wrote assignments for wizards, for ghosts, and for westerns. I wrote one for autobiography and another for creation stories. Each year I tried to come up with something new. In order to connect assignments to the mandatory inquiry project, I asked students to write stories that explored the subtler aspects of their research topics using the genres metaphorically. What I found was that students not only improved their problem-solving capabilities by writing stories that met the rigorous demands of the assignment, but that doing the assignment also expanded their thinking about their research project.

Unexpected Parallels in Engineering

One particular assignment opened opportunities to collaborate with Engineering. This one focused on robot narratives. I usually introduced this assignment by reviewing the conventions of robot-based science fiction, examining their common themes and questions and the manner in which they brought the author's creative faculties to the task of envisioning solutions to some of our more pressing problems. Robot stories tend to address human anxieties and hopes concerning technology—as well as the concepts of programming and learning and the relationship between creator and creation. We would examine some of the more straightforward approaches to the robot narrative in texts such as *The Matrix, Terminator, Wall-e* or *AI*, and then we would look at less obvious iterations of the robot concept in stories such as *Lord of the Rings, Pinocchio,* and *Genesis*. We used the robot concept as a metaphor to look at the world around us, to examine the past, and to predict the future. Figure 3 displays the list of rules students were required to follow.

What I like about this assignment is the manner in which it forces a question to appear, not just once but twice. My hope is that the question will parallel some of the questions students are asking in their research projects. The enormous burden of addressing big-ticket items like war and bigotry, law, and what it means to be human forces students to think deeply about the subject of their research. They have to consider the grand narratives of history, the concept of translation, and they have to engage the issue of intelligence. All of these issues can be addressed simply by writing about some sort of killer robot wreaking havoc on humanity, but interestingly many students avoid the robot completely and instead use the genre metaphorically to write about the everyday.

Following is an early draft of one student's response to the robot assignment. As a military veteran and prolific autobiographer, Benjamin "Jet" Garte chose to use the rules to write about his life in the Marine Corps. What I like most about his story is the way he is able to make the rules feel necessary and natural. While the parenthetical notations are there to exhibit how the student followed the rules, the rules do not interrupt the flow of his story. His writing is fluid in spite of the 40+ requirements; the characterizations are vivid; the experience is conveyed clearly and in a voice that is energetic and engaged with its own material. Such a display demonstrates "high-order" writing skills, demonstrating a complex understanding of the genre itself while using it to examine a subject that he has been researching for his semester-long inquiry project, the Marines. The student was kind enough to allow me to include this early draft of his story in this essay so that readers might see how these rules could be used to construct a narrative that is fluid and spontaneous

Pop Culture as Social Critique: "Robots"

<u>Requirements and Specifications:</u>

Write something that follows the rules below in less than 2000 words. Note all rules in your copies of the paper using the parenthetical codes listed below. Check off this rule sheet and attach to the front of my copy, along with a cover letter reflecting on your process and the product of your work, posing at least three questions you would like feedback on during workshop discussion. Bring 23 copies total to the start of class on the due date. Your paper must be formatted according to the syllabus. Papers that do not follow all rules cannot receive a passing grade. Papers that follow all the rules cannot receive a grade below a C.

1. ____ Do not use the word or any derivative of the word "robots" anywhere in your paper.
2. ____ Primarily present tense narration
3. ____ Addresses the concept of creation (CREATION)
4. ____ Addresses the concept of intelligence (INTELLIGENCE)
5. ____ An emotion must be projected onto an inanimate object (OBJ)
6. ____ The inanimate object must cause the main character to experience a problem (PROB)
7. ____ Defines what it means to be human (HUMAN)
8. ____ Makes reference to a war (WAR)
9. ____ Makes three references to law (LAW 1-3)
10. ____ Features a debate (DEBATE)
11. ____ Addresses bigotry (BIGOT)
12. ____ Includes two warnings (WARNING 1-2)

Fig. 3 "'Robots': requirements and specifications" (Crickenberger, 2015b)

13. ____ Provides a brief history of the human world (HIST)
14. ____ Contains a machine that malfunctions (MACHINE)
15. ____ Includes an ambitious character (AMB)
16. ____ Includes a nervous character (NERV)
17. ____ Includes a detailed description of something naturally beautiful (BEAUTY)
18. ____ Begins and end with the same question. (QUESTION 1-2)
19. ____ Includes a detailed description of a robot (DESCRIPTION)
20. ____ Translation (TRANSLATION)
21. ____ A lapse in time (LAPSE)
22. ____ Engages concept of autonomy (AUTO)
23. ____ Five references to lighting or weather (L/W 1-5)
24. ____ A detailed description of someone's hands (HANDS)
25. ____ Five references to smell (SMELL 1-5)
26. ____ Five references to sound (SOUND 1-5)
27. ____ Five references to a texture (TEXT 1-5)

Fig. 3 (continued)

seeming. He also allowed me to include his cover letter in which he was required to reflect upon the writing process. See Benjamin Garte's first attempt at completing the assignment in Fig. 4.

Benjamin's exploration of the machine-like quality of the human body and the rule-based living that goes hand in hand with life as a Marine raises all sorts of issues for discussion. In the narrative, we are given a close look at the kind of training Marines endure and we get a sense of how this individual found inspiration and motivation to improve. Benjamin went on to revise this story for his final portfolio where it finally appeared as part of a five-essay collection of autobiographical narratives that each corresponded with a different genre studied in our class.

I was quite often happy with my students' attempts at solving the problems presented by the assignments, and strangely, unlike many of my colleagues, I did *not* find engineering students to be "the worst writers in my class," quite the contrary. In fact, I noticed pretty early on that my engineering students seemed much less intimidated by the task than students in other majors. They would jump in quite readily, happy to work within the conventions of science fiction, creating their own robot stories, exploring their own anxieties about technology or their own questions of existence. Many drew on knowledge gained from their engineering classes, writing about those issues directly. Others would write about their lives, exploring the robot-"esqueness" of the everyday. More often than not, I was struck by the vivid

<u>Cover Letter</u>

This paper connects to my project directly, the Marine Corps. This is a story about a Marine working with another Marine to get better at his PFT for promotion. The process I used for this writing involved a lot of class work. When Dr. Crickenberger first introduced the creative challenge to the class, she had us take out our sketch books. While she was going through all of the rules, I was writing possibilities in my sketch book. My sketch book is mostly writing rather than any illustrations. My original idea is the same one I wrote about, a Marine that is amazing at PT, a "machine" in a sense. I looked over these notes a couple times over the weekend but didn't put anything down on paper. I didn't start writing until I got into class at about 4:40pm on Monday (about 20 minutes before class started). Hopefully without incriminating myself, I will admit that I did continue writing until about 5:10pm, as we were going over the rules again. I couldn't help it as I kept getting good ideas for the next couple paragraphs. I finished articulating all of my thoughts and got back to class work. After making my way home and eating some baked chicken I started on the story again. I finished it up and tweaked it a little bit. I couldn't seem to find a way to fit all of the rules in there so either I am going to treat this as just a regular assignment or I will have to strongly revise my paper. Other than following the rules in general, one of the questions I would like to ask in class would be about the question rule in particular. My question is worded differently between the beginning and the end. I would also like to know if it has to be the beginning of the paper or can it be the beginning of a paragraph?

Fig. 4 An early draft of Benjamin Garte's "PT Animal" in which is noted all Robot Rules (Garte, 2015)

PT Animal

By Benjamin Garte

"How are you such a beast? (QUESTION 1)" I ask Porter as I meet him outside of the barracks near the not-so-broken, spastic, flashing light. (L/W1) The kind found in horror movies, it even has a feint buzz that can be heard from up close (SOUND 1).

My name is Lance Corporal Jet Coburn, but most of the Marines call me Jet, because they all suck at spelling. I would say the most interesting thing about me are my 9 and a half fingers. One was crushed by a TV when I was a toddler, leaving only a rough nub where the rest of my pointer finger use be (HANDS/TEXT1). I am what you would consider an "average Marine", minus half a finger, standing at five foot six inches tall and weighing in at a whopping 165 pounds. I am completely under-motivated when it comes to PT for the most part, but I'm willing to put forth the effort with the right reward. Seriously, if all that the Marine Corps requires of me are the two tests a year, it would be the only time anyone might see me run. However, in this case, there is a reward in the form of a promotion to Corporal, so I will be giving it my absolute best effort for this upcoming Physical Fitness Test (PFT) (AMB). To get promoted I have to beat out other Marines of my rank, making it a very serious battle to be the best (WAR).

Fig. 4 (continued)

imaginations my engineering majors had and their ability to visually construct a machine with language. It was through these assignments that I started to see how creative the field of engineering really is.

In 2008, after one of my students who worked as an administrative assistant for the Engineering lab took my assignment to the head of writing-intensive classes in the Department of Mechanical Engineering, I was offered a job as a technical writing advisor. "Your assignments are just like the assignments we give in engineering," they said to me as they showed me the three-page checklist in the back of their *Formal Technical Reporting Guidelines* (The William States Lee College of Engineering, 2015). This checklist can be seen in Fig. 5.

To get better I work out with a friend and colleague of mine, Corporal Phillip Porter. Porter is the perfect example of a stud. He stands five foot nine inches tall, very muscular yet lean, chiseled and cut almost as if the Greek God's were modeled after him (TEXT 2). Not only can he destroy a workout, he has the answer to every question I have been able to think up, from information about a helicopter's up-draft to things that a bank requires to attain a low interest mortgage, he knows it all. (INTELLIGENCE) Porter doesn't stop at physical fitness or knowledge, he is extremely serious about everything to do with his job, the Marine Corps (DESCRIPTION). I think he will be happy to prepare me for the PFT next month, which would be true were he able to register any kind of emotion.

The PFT is the standardized test taken by Marines during the spring time to show performance in three different workouts. There is a three-mile timed run, crunches, and pull ups. To get a perfect score, which I am aiming for (AMB), I will have to run three miles in eighteen minutes or less, do one-hundred crunches in under two minutes, as well as twenty pull ups consecutively without dropping off the bar. This PFT will determine whether or not I get promoted next quarter to the rank of Corporal, I need to keep my eye on the prize.

Porter and I start the workout with the daily sixteen, the exercises Marine Corps requires (LAW1). The daily sixteen is a group of warm-up exercises that help to get the blood flowing and muscles

Fig. 4 (continued)

This checklist resides at the back of the *Formal Technical Reporting Guidelines* that are currently used in UNCC's Department of Mechanical Engineering for writing intensive labs. While the content here is obviously very different from that of my assignments, the structure of the assignment is almost identical, a list of specifications and requirements designed to force thorough engagement with experimental data. It was upon seeing this similarity that I began to understand ways that the work I was doing in First-Year Writing could transfer to STEM disciplines as a means of gaining experience with design thinking.

loose for a workout. We start with side-straddle-hops, also known as "jumping jacks (TRANSLATION)." After completing the remaining 15 exercises we move to stretches. Now, stretches are not "required" by the Marine Corps, however, they are mandatory prior to a run (LAW 2). Porter and I make our way to the gravel road (TEXT3) after finishing the stretches.

I get to the road and begin to take in my surroundings. The time, I figure, is about 6:00am in the morning. I can smell the dew still residing on the grass surrounding the road (SMELL 1). It is still mostly dark outside, but I can see the light from the sun begin to come over the horizon, one of the most gorgeous sights in nature. (BEAUTY) I lose focus and start ponder to myself how this event has been happening for millions of years on Earth. (L/W 2/ HIST). Also, how according to a "clock" theory, humanity has been around for less than a minute of the 24 hours that represent those millions of years.

It is still dark outside (L/W 3), so Porter and I are wearing our glow belts because it is a requirement by Marine Corps Order P1020.34G (LAW 3). The order states that glow belts must be worn if a Marine is outside from sun-down until 8am. Glow belts are a reflective (L/W 4), usually plastic (TEXT 4), belt that goes around the waste, making the wearer visible to other people or vehicles (WARNING 1) that may be driving about. The glow belts are probably the most uncomfortable thing to wear when trying to run multiple miles. They rub

Fig. 4 (continued)

As one who has graded these lab reports for many years since, I have had the opportunity to see my own freshmen again as they submit their reports for English grading in their senior-year *Thermal Fluids Lab*. I find students who did not take my freshman writing class are often not as adequately prepared to follow such a complex set of rules. They have difficulty stitching all of the required bits of information together into something cohesive and fluid. My students, on the contrary, often do very well with the technical reporting assignments, and I am convinced it is because they have had experience working creatively within constraints and approaching the task of writing as an occasion for design. They are not intimidated by complexity and they have developed the ability to preserve their voice while still tenaciously checking the required boxes.

against your clothes, causing very noticeable discomfort, which is bothersome when trying to focus on anything, let alone running.

Porter asks me "Are you ready for the run?" I reply "I am as ready as I will ever be, although I am a little nervous (NERV) about being timed for this one". Porter says "Don't think about the timer, just keep pace with me and you will easily get a perfect score". I nod my head and we make our way to the start point in the form of a fading line of white paint (TEXT 5) on the road. Porter looks down at his watch, "Get ready, 30 seconds until we start". I inhale through my nose, catching a whiff of the surrounding pine trees (SMELL 2), I then exhale slowly, focusing my eyes on the road in front. He counts down "Three, two, one, Go!" I hear the beep of Porter's watch as he starts the timer. (SOUND 2)

As we run I try to keep pace with Porter, but he is an animal and begins to move farther and farther ahead. He breaks through the wind without stopping, taking long strides that make almost no sound (SOUND 3). Breathing calmly, he runs like this doesn't even phase him, shouting "KEEP PUSHING" to me as he disappears in to the darkness ahead (DESCRIPTION/ L/W 5).

I have no idea how long I have been running, this seems to happen to me often without even thinking about it. (LAPSE/AUTO) Ahead I see Porter's glow belt static in the distance, I know the finish line is close. "PUSH IT BRO, YOU'RE ALMOST DONE!" Porter

Fig. 4 (continued)

Exploration and Design Thinking

In spring 2015, I presented these assignments to writing and STEM faculty at NC State University's *NC Symposium on the Teaching of Writing* (Crickenberger, 2002) where I first publically discussed my assignments in regards to the issue of *transfer* between writing classes and STEM fields. In order to prepare the presentation for a STEM audience, I had to research the language of design thinking. I began by looking into the field of writing studies where I encountered Leverenz (Leverenz, 2014) and the concept of the "wicked problem" as a pedagogical tool. "Wicked problems,"

yells during my last stretch. I give everything I have left. As I cross the finish line, completely out of breath, I hear another beep (SOUND 4) from his watch. "Time" Porter says. Before he can tell me my time I run over to the grass and throw up, something that I am experiencing for the first time, the smell was terrible (MACHINE/SMELL 3). I force myself to stand back up, with the taste of puke still in my mouth, I walk over to where Porter is standing.

I look at Porter; he is just standing on the side of the road with his arms crossed. I can't hear him breathing heavily at all (SOUND 5), nor is he sweating. It is almost unnoticeable that he has run at all in contrast to me, I am dripping sweat. To make things worse, I forgot to put on deodorant this morning which is causing my body to produce the most putrid odor, it could kill a small animal (SMELL 4).

I ask him "What was my time?" Porter tells me "eighteen minutes and two seconds, you will have to push harder next time." "Two seconds? Really? You can't just give me the two seconds for a perfect score?" (BIGOT) I ask, growing angry. "No." Porter states very calmly "You have to earn that time. I am not going to give you two seconds. You won't want to try harder next time and you will probably do just as bad, not getting you that promotion. That and it is against the rules (LAW 4) to cheat."(DEBATE/WARNING 2/CREATION) I shoot back at him "Well, I am not like you, the God damn PT stud. I can't just run three miles in ten freaking minutes. I

Fig. 4 (continued)

a term coined by architect Horst Willhelm Jakob Rittel and defined by Churchman in 1967 in the context of Management Science, are those problems which are confused by containing too many options and too many conflicting values (Churchman, 1967). Churchman takes the time to look at these problems in a moral context, arguing that managers tend to smooth over these sorts of problems without ever really getting a handle on them, often creating "solutions" that are worse than the original problems. Writing poses a situation similar to management science in its wicked nature. Often, students will write the easy thing, the thing they've heard before, or simply rephrase something they found on the internet. But writing is never easy if its purpose is to

work hard for what I want. I wasn't created the perfect Marine like you were. I am just a regular person." (HUMAN).

Ignoring my anger completely, Porter walks over to the pull up bar. His calmness rubs off on to me, and I settle down and follow close behind. Porter gets up on the bar and pumps out twenty perfect pull ups without the slightest hint of effort. I have no idea how he does it. I jump up on the bar after him. I start pumping them out. My first seventeen are acceptable, but then I begin to struggle. I barely push out my last three for a total of twenty. "Good job!" Porter says unconvincingly "Let's knock out those crunches."

We make our way over to the grass. It must have been cut while we were running because there are blades of grass (TEXT 6) everywhere along the sidewalk, as well as the overwhelming smell of freshly cut grass (SMELL 5). I lay down on my back and get in the crunch position. I go first, with Porter holding my legs, doing my hundred crunches in just under ninety seconds. Porter does his crunches, without my assistance, completing his hundred in about half the time that it took me.

"Good job today Jet, you just need to work on that run and you will be ready for PFT and hopefully that promotion." Porter tells me. I thank him for the help and set a date for next training session, I start to walk back to my room when I hear Porter ask "Hey, I am about to go work out, do you want to join me?" to which I reply, "**** no, I am completely dead right now. You are such a machine. Seriously how do you do it? (QUESTION 2)"

Fig. 4 (continued)

both expand and share knowledge as it is in academia. Leverenz, drawing on the work of Richard Buchanan who brought the concept of the wicked problem into the realm of Rhetoric and Philosophy by encouraging mindfulness of the arguments technologies make implicitly through their design, recommends in her 2014 article, "Design Thinking and the Wicked Problem of Teaching Writing" that writing teachers use design thinking in their classes by teaching the writing process as one of

Checklist for Formal Technical Reports

Using this checklist will help to ensure a professional laboratory report. However, it is recommended that you read and follow the Formal Technical Report Guidelines for complete requirements.

Report Contents

Title Page

__ Report includes a title page

__ Title of report is descriptive

__ Name of author is highlighted/underlined

__ Names of laboratory partners included

__ Name of class included

__ Date experiment was performed included

Table of Contents

__ Report includes a table of contents

__ Major sections of report and corresponding pages listed

__ Subsections listed (if applicable)

__ Tabbed leader line used to connect headings with respective pages

Abstract

__ Report includes an abstract

__ Abstract is approximately 200-300 words and one paragraph

__ Abstract written in past tense, passive voice

__ Abstract summarizes objectives, methodology, main findings, and conclusions

__ Abstract cites key statistics and data

Objective

__ Report includes an objective.

__ Objective is one or two concise sentences.

__ Objective describes purpose and motivation of the experiment.

__ Objective is written in passive voice.

Fig. 5 A copy of the checklist we use to evaluate technical reports in our writing intensive labs. This checklist resides at the back of the forty-page *Formal Technical Reporting Guidelines* that we require our students to follow when writing their reports (The William States Lee College of Engineering, 2015)

Introduction

__ Report includes an introduction.

__ Introduction gives the scientific background of the experiment and explains its context.

__ Introduction introduces relevant theoretical models and key equations.

__ Introduction is one to two pages in length.

__ Introduction is written in passive voice.

Theory

__ All equations that are used for analysis or modeling should be introduced, presented, and explained in this section.

__ Analysis methods such as least squares are detailed.

__ Analysis procedures or algorithms are described.

__ It should be written in the present tense.

Equipment and Procedure

__ Equipment and procedure section is written in paragraph form.

__ Equipment description includes model/serial numbers, date of calibration, etc.

Fig. 5 (continued)

design, by using wicked problems as writing assignments, and by encouraging collaboration and experimentation through the act of prototyping (Marback, 2009). She argues that adding complexity to assignments can enhance their effectiveness in student learning. This is a notion growing in popularity as written texts evolve to incorporate more complexity in the form of multimodal compositional elements in electronic and networked contexts.

As I attempted to connect this way of looking at writing as design to the field of engineering, a field that is primarily focused on design problems, I encountered Hatchuel & Weil's "C–K Theory" (2003) (Crickenberger, 2015c) which draws upon the mathematical principle of "forcing" in set theory as described by Paul Cohen in 1963. Although the authors of C–K Theory do not mention wicked problems, the manner in which they use "forcing" in their model to generate creativity in the design process speaks to the manner in which my writing assignments assist students in the writing process by forcing them to overcome fixation effects that prevent them from delving more deeply into their subject matter. The forcing is done

__ Experimental setup is described in detail.

__ Conditions under which experiment was performed are described. NOTE: describe what actually took place in the experiment, as opposed to cutting and pasting what was supposed to take place as proscribed in the lab instructions.

__ Diagrams, sketches, and photographs of equipment and material are included.

__ Procedure is detailed enough to allow repeatability of experiment.

__ Equipment and procedure is written in past tense, passive voice.

__ Laboratory manual instructions are cited but NOT repeated.

Results and Discussion

__ Results include figures and tables.

__ All figures and tables are introduced, referenced and discussed in this section. Be sure to explain the general relevance and significant attributes of each: read the figures and tables to your readers.

__ Results and discussion section is written in present tense, passive voice.

__ Figures and tables are placed near the discussion that refers to them.

__ Raw data and extensive tables are placed in the Appendices.

__ Results are thoroughly explained, analyzed, and interpreted.

__ Key trends and relationships are explained.

__ Experimental results are compared to literature or expected values.

Fig. 5 (continued)

through the introduction of "crazy concepts" in the design process. Crazy concepts are proposed solutions to a design problem that are not limited by the actual, the attainable, or even the real. These proposed solutions are never intended to be implemented, due to excessive time involvement, expense, or because they are simply beyond our capabilities as humans, but they are considered in the design process because considering them helps us move beyond what we know. Figure 6 displays a visual rendering of how C–K Theory works.

In Fig. 6, Hatchuel and Weil's theory is represented in terms of two spaces of thought—the "space of concepts" and the "space of knowledge." The model forces its user to move back and forth between previously acquired knowledge and potential

__ Possible reasons for discrepancies are given.

__ Comparison with similar investigations (i.e. other laboratory groups) if required.

Conclusions

__ Conclusions section is written in past tense, passive voice.

__ Overall conclusions relate directly to the original objectives of the experiment.

__ Main objectives are restated along with to what degree they were achieved.

__ Principles, laws, and theory that were validated are explained.

__ Possible sources of error are explained.

__ Applications of results are given and their importance to engineering is explained.

References

__ Chicago style for the sciences is used for all in text and end of report referencing.

__ Laboratory manual, textbook, and other key sources are accurately referenced.

Appendix

__ All hand calculations, raw data, extensive tables, etc. are included in Appendix.

Fig. 5 (continued)

applications for this knowledge in the creation of a design solution. Again, the crazy concept is never intended to be realized as a solution to the design problem, only to expand the knowledge base upon which a more sensible solution might be created.

In my presentation, I discussed a few of my creative writing assignments in the context of the writing my students have to do in writing-intensive labs. The "robot" or "alien" requirements work much in the same way as Hatchuel and Weil's crazy concept, insofar as these metaphorical figures function, not as ends in themselves, but as means to expanding the realm of possibilities from which students draw the substance of their stories and the manner in which they connect these endeavors to

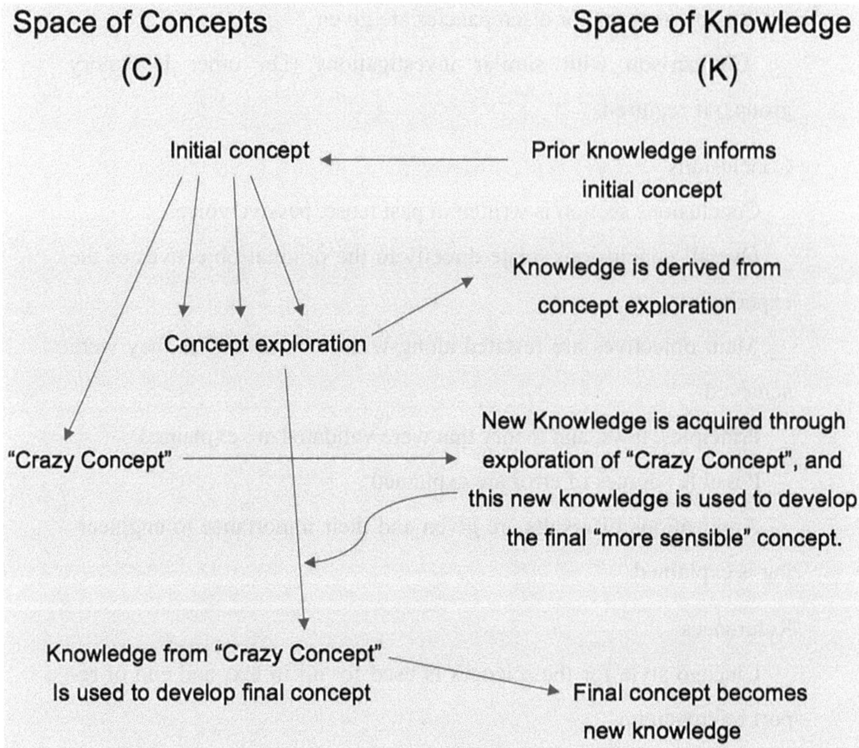

Fig. 6 Visual representation of "C–K theory" (Armand & Weil, 2003)

their semester-long inquiry projects. I was surprised at how readily STEM faculty were willing to accept this approach as something that would be useful to their students: "systems thinking" some called it—"design," they would nod. "It's what we do every day." "It's an algorithm!" The more I talked to professors in design-based fields, the better able I was to bring back those applications to my own institution.

Since this presentation, my experience with the College of Engineering led to my being asked to teach a special section of our First-Year Writing in the Engineering Learning Community in the Fall of 2015, where I was able to work more purposefully with unpacking the elements of design thinking inherent in my own assignments so that engineering students might recognize their relevance to the work they will be expected to do as engineers. I found that these engineering students were not only able to discover their own creative capabilities in the realm of science fiction, but were able to use those creative writing assignments to inform and improve upon the research and inquiry work they were doing in our program's required multimodal research project. Examining their research topics through the metaphorical lens of science fiction opened new questions and new kinds of thinking about science and technology, deepening students' engagement with their subject matter. They also began to see how crazy concepts can inform serious research by taking it in a new direction and thus opening up new pathways to previously undiscovered solutions. Many also learned that they liked storytelling and that narratives that create a clear picture and are tuned to a specific audience are an integral part of communication in engineering, as in any field.

Following are two stories written by engineering majors who participated in this particular section of First-Year Writing. The first story, "The Fearless Soldier," by Samuel Faries explores the issue of error and short-sighted thinking in relation to the weapons industry and the possibility of a human-designed weapon of mass destruction that develops the ability to make decisions autonomously. I must emphasize that this is an early draft of the paper and that the story itself was later refined in a draft that was not required to follow the list of rules. I chose to use these early drafts because the rules are more visible here and because I think teachers might benefit from seeing the variety of writing-related issues that might be addressed when working with an assignment such as this one. See Samuel Faries' first attempt at completing the assignment in Fig. 7.

In this story, the student took the opportunity to employ much of what he had learned in researching weapons development. Specific questions concerning reasons for pursuing such a career are addressed as well as a few engineering concepts that the student had explored in his first year of engineering training. In his story, he actually designs a weapon, but he also designs a context for that weapon, one populated by human beings and governed by regulations. I find it interesting that the student begins and ends with the question of past decisions controlling the future.

Cover Letter

This story connects directly with my genre of topic being weapons and it connects to me because it is what I want to do with my career. (More successfully of course!) This is a story of how weapons and the engineering behind them can have such an impact on the entire world. In the correct hands they can be tools that save the world. However, in the wrong hands they can destroy it. I am pursuing a mechanical engineering degree to become a weapons engineer in hopes that I can impact the battlefield in a great way. My process for writing this story was to look at the rules and find ways to meet these rules. I then went on to find ways to put these ideas together as smoothly as I could. I didn't meet all the rules because I was pressed for time but I did intend

Fig. 7 An early draft of an engineering student's attempt at the "Robots" assignment. This student was required to connect the concept of robots to the topic of his semester-long inquiry project, about which he chose to research the field of weapons engineering (Faries, 2015)

on this being one of my final creative challenges.

The Fearless Soldier

By Samuel Faries

Do you think the past controls the future? (QUESTION 1)

I am currently reading a very old book called *The Machines of World War I*. When you open it, it creaks and smells like dirt and dust with very smooth pages. (SMELL 1) (SOUND 1) (TEXT 1) It is very interesting to me, the history this book holds, and it contains possibly the one idea that could influence me in a great way. It talks about all the new machines used in World War I that had never been seen before. (WAR) New machines, machines of great power that could save lives by putting an end to many.

I am a weapons engineer for the United States Army, and I am reading this book in hopes of it inspiring me to create my own machine of war in order to provide protection to the United States and to those who are risking their lives to protect our freedoms. You see, necessity is the mother of all inventions. Without the need or want for something, it will not be invented. The war efforts in the past have been hindered because of soldiers dying in combat, so much so that, at some points, people were used as equipment, like in the Battles of Verdun and the Somme—solely Battles of attrition. On those dim and dreadful mornings, the soldiers quickly lost interest and morale and started

Fig. 7 (continued)

His rumination on this subject demonstrates that he is considering issues of ethics and responsibility in the design field he has chosen, but he is considering them through the faculty of imagination as it is overlaid onto his education and experience in the form of creative writing.

The next story included here is "What Is It?" by engineering student Nicholas Fritts; it examines the subject of his research project on experimental diseases. At the point that he constructed this story about an advanced medical robot designed to locate infected humans, he was about half-way through the research phase of his semester-long inquiry project on diseases. Again, we see the imagination come into play as the student engages the limitless world of science fiction in order to visual-

abandoning their positions. (HIST) (LIGHTING/WEATHER 1)

Now, at this moment in time, there is very much a need for someone who can stand up for what they believe in, no matter what the cost, someone who will not turn away from anything, no matter the situation. While reading my book, I feel someone grab me from behind and put a dark, coarse bag over my head. (LIGHTING 2) (TEXT 2) They are very stern with their commands.

"Stand up," a voice shouts at me from across the room. "Come with us

As If I have a choice. One of the men puts me in a vehicle that smells of gun powder and machine oil and drives for what seems like days. (SMELL 2) Finally, we get to the place that they desire. We leave the vehicle and walk through the pouring rain.

Once inside, they lift the bag from my head. (LIGHTING/WEATHER 3).

"Congratulations, your dedication to the Army and your knowledge of weapons has landed you a position on the MAARS team."

MAARS is the Modular Advanced Armed Robotic System. (CREATION) Looking around, I see just a few other engineers that I know were put through the same situation that I was.

"Sorry about that fellas, but you are on a classified team working with the highest technology our military has to offer so no one can

Fig. 7 (continued)

ize potentially disastrous scenarios surrounding the research field that he has chosen. I love the way the narrative is populated with engineering terminology. Again, it is an early draft, and the student was given the opportunity to revise this story without regard to the requirements and specifications, so there was room for him to expand on his ideas in a version that was not limited by word count. See Nicholas Fritts' first attempt at the assignment in Fig. 8.

In both stories, the students used the crazy concepts posed by the wicked problem of a rule-based writing challenge to advance their thinking, not only about the genre they were working in, but about the research they had been working on all

know where you are or what you are doing."

As I look around at the men that brought me here, one throws a packet at me and it read "WARNING!!! CLASIFIED INFORMATION DO NOT DISCLOSE." It felt like a brick and smelled like nothing but a job for me to do. (WARNING 1) (SMELL 2) (TEXT 3)

Right now I am being told from my superior officers that there is a need for a weapon on the battlefield to keep men out of harm's way while maintaining their ability to engage the enemy. My weapon must comply with the Geneva Convention and my superiors' orders:

> *Law 1: the weapon cannot succeed by the employment of arms which uselessly aggravate the sufferings of disabled men or render their death inevitable.* (LAW 1)
>
> *Law 2: the weapon must abstain from the use of bullets which expand or flatten easily in the human body, such as bullets with a hard envelope that does not entirely cover the core or is pierced with incisions.* (LAW 2)
>
> *Law 3: weapons which by their nature are incapable of being directed specifically against military objectives, and therefore that put noncombatants at equivalent risk, are forbidden due to their indiscriminate*

Fig. 7 (continued)

semester. By engaging their subjects imaginatively, these students were able to produce coherent narratives that helped them to think through the material they had been working with, while at the same time creating an opportunity for other students in the class to engage the subjects of weapons design and experimental diseases in metaphorical and nonconventional ways through our workshop discussions. Both students had the opportunity to share these stories and to converse about them as creative works of art with their teacher and peers, opening up new pathways for their research, new questions to explore, and new perspectives on the issues that they had been studying.

effect. (LAW 3)

Before I begin the designing process, I first want to get an idea of how the machine will move, so I ask the other engineers what they think. "Tracks" is the approach quickly settled upon because they are very versatile and do not violate sound discipline when they move. (SOUND 2) They can go over anything, even up and down stairs, and provide a stable platform for accurate shooting. After a little more talking, we go through and chose all of the aspects of the MAAR system. It will move with tracks, have a smooth armored housing for the electronics, and have an attachment point on top to attach multiple things as needed such as cameras for surveillance or many different weapons of choice. (DESCRIPTION) (TEXT 4) As I go through the project, the excitement of being able to provide soldiers with a weapon that can keep them completely out of harm is overwhelming me. As a matter of fact, I can hear the sound of the crowd when I am named top engineer in the United States. (SOUND 3)

I instantly go on past my team members that are beating metal and burning circuit boards, to think more on the weapon itself while they stay behind and look at the small things. (AMB)(SOUND 4)(SMELL 4) It is very hard for me to come up with something that has never been seen or ever considered for use on the battlefield, but with time and patience, I make a computer program that targets the enemy.

Fig. 7 (continued)

How I Use These Assignments in First-Year Writing

Because I have such a large number of these wicked assignments engaging so many different genres, I often give students several options to choose from. Each genre carries with it its own traditions, its own vocabulary of images and stock moments, and a somewhat consistent metaphorical significance. Students begin to see these trends once they start working within their selected genre, developing yet another skill that can be transposed onto the manner in which they engage the conventions of genre-specific tasks in engineering.

* * *

When it is finally finished, we load the program that I made onto the drive located at the heart of the machine—to keep it safe from bullet fire. We put the final touches on the weapon, such as the message, "WARNING FRONT TOWARD ENEMY," because this was the normal warning for our military. (WARNING 2).

"Are you sure your program will work? If it doesn't it will kill innocent people, it could even turn and kill our troops," one of my team members asks in the final review. (NERV)

"Are you serious?" I say, "Turn and kill our troops? It doesn't have its own mind or way of thinking. It simply does what I programed it to do." (INTELEGENCE)

We put it through a test trial before sending it into combat. Walking to the trials I can feel myself getting livid with every step because we are having to go through all of this just because they questioned my programing. (BIGOTRY) After seeing the MAAR System work flawlessly they understood that it would work the way I programed it to. Right before we ship it off, we stand there looking at it like as if it was a sunset over a beautiful mountain range. (BEAUTY) (LIGHTING / WEATHER 4) We are very pleased with our many months of work. (LAPSE) I finally get to go home, the sound and smell of home relaxed me from my many months of stressful work, now that my mission is complete. (SOUND 5) (SOUND 5) The next

Fig. 7 (continued)

To start, I go through all the rules with them one at a time, asking them to jot down any ideas that come to mind. As I introduce each assignment, I give them some background on the genre, showing them examples and having them do some exploratory writing and small group discussion. As I mentioned before, in recent years, our program has required a semester-long inquiry project, so I have the students add one more rule to the list: *You must connect your narrative to your inquiry project in a way that you explain in a cover letter of at least 100 words.*

morning, the MAARS System is deployed overseas on a team with a trained mechanic to monitor the system. It is reported to be working fine until it runs over an I.E.D. Shortly after that, the mechanic reads a code that translates to "ENGAGE ALL TARGETS." By the time he starts screaming telling everyone to run, the system turns and starts killing everyone in sight. (TRANSLATION) (AUTO) (MACHINE).

When this news comes back, the day is dark and gloomy, and I find myself giving my hands a hollow stare. (LIGHTING / WEATHER 5) Looking at them, they are coarse with wrinkles and age, creases full of the pride that caused all of this. (TEXT 5) I can't believe I made all of this happen with my own two hands. (HANDS) When this news reaches the public, everyone explodes with emotions and bombards me with questions.

"Do you think it is humane to let a machine decide who the enemy is?"

"Do you think it is okay to stand behind a wall and kill people that have no chance at defending themselves against a machine?" (DEBATE)

Now the Army mandates that every AWS (Autonomous Weapon System) must be used for defense only—not offence—and the weapon systems that are used for offence must be controlled by a human. The Army defines a human to be someone that is able to make decisions in real time, in a way that is similar to the vast majority of the population and is able to be punished when the wrong decisions are made. (HUMAN)

Do you think the past will control our future? (QUESTION 2)

Fig. 7 (continued)

Students do quite a bit of thinking in order to choose their genre, searching for parallels between the subject of their research project and the rules they have to follow. For the first draft, they are only required to do the cover letter and write at least 750 words concerning their chosen genre. Some students use this as a freewrite, while others follow all or nearly all the rules. These drafts are very low stakes, making up only 5 % of their grade. They must read these drafts aloud in groups of three or more discussing what was best and what could be built upon in their next draft.

Of course the early drafts of these assignments are seldom flawless. Every semester, I have a handful of students who write excellent first drafts, but most are clumsy, sloppy, and awkward on the first round. Students still haven't figured out how much freedom they have to interpret the rules and they haven't come to terms with how soft and permissible language can make things.

Cover Letter

. . . My inquiry project is about experimental diseases and the possibility and likelihood of them somehow being released from research lab(s). I start off with a fictional story about a person living in the world after such events. . . . When writing a piece like this, I like to wait until later at night when I am very tired. Often I think of interesting ideas but when I'm fully awake I often stray away from them. When I am tired I just write them down. I use a lot of pop culture for my inspiration. . . .

What is it?

by Nicholas Fritts

Could it have happened to me? (QUESTION 1)

The shadow moves closer. There is nowhere to go. It is has the only exit blocked. How had it gotten in here? And what is it doing? Peering outside past it, all I see is a sheet of rain (L/W 1). Lightning is striking erratically (L/W 2). I hear the roar of thunder followed by clanks as the shadow approaches (SOUND 1-2) (L/W 3).

Man, would I like to get out of here, away from this thing.

I imagine being back in the old world. I long to reach into a fridge filled with food again. In my mind, I pull eggs, milk, butter, bread, and bacon out and begin to cook. A lovely aroma fills the kitchen (SMELL 1). First the scent of cooking eggs (SMELL 2). Then, suc-

Fig. 8 An early draft of an engineering student's attempt at the "Robots" assignment. This student was required to connect the concept of robots to his chosen semester-long inquiry project which concerned weapons engineering (Fritts, 2015)

culent bacon being masterfully prepared in a sizzling pan (SMELL 3). I smell the bread being toasted (SMELL 4) (LAPSE). All of a sudden I snap back to reality with the approaching figure. My nose is sent into shock as it is brought from the delicious smelling kitchen to the musty concrete place I have come to call home. (SMELL 5)

It's all made of metal, but it walks like a man and even has a voice like a man. "Stay still," its' voice says to me (SOUND 3). I lunge forward in an attempt to get around it and out the door. A cold steel hand grabs my arm and holds me still (TEXT 1). The hand is smooth but firm (TEXT 2). I can feel the strength of the figure's grasp on my arm. I look down at it. It is cold and lifeless but at the same time looks extremely similar to mine with fingers, five to be exact, tendons, and joints. Looking over at the other hand my heart drops. This isn't a hand. It's a needle (DESCRIPTION) (HANDS). I watch helplessly as the cold steely metal pierces through my course skin (TEXT 3, 4, & 5). I feel a prick and watch as blood is drained from my arm.

I'm going to die...

To my surprise, the needle is removed as quickly as it was inserted. A bandage is placed over the wound and it begin to leave. I grab its' shoulder and ask, "What did you do to me?" As I jerk the figure, a wire comes loose and it crashes to the floor making a deafening clank (MACHINE) (SOUND 4).

Old reruns of *Family Matters* pop into my head and I just laugh

Fig. 8 (continued)

On the day of the first roundtable workshop, students arrive with copies of their stories for the whole class. These drafts must meet all requirements and specifications and must be presented as cleanly formatted hard copy. We move our chairs out into a big circle and pass the stacks of stories around, and then I have everyone put the stack of stories in alphabetical order by first name: that stack becomes our textbook for the next 3 or 4 weeks. Each class, three or four students read their stories to us and then we spend about 15 or 20 min discussing them. I take notes on my copy of each student's paper and then write a bit more feedback later before returning it to the student. I know some professors who think having students read their

as I say, "Did iii dooo thaaat?"

"Yes, you did do that. Now plug the wire back in so our equipment can get back to work," comes a stern voice (AMB) (SOUND 5).

"If you tell me what the heck is going on, I'll plug the wire back in."

The voice emanates from an invisible sound system, electronic, cold: "It was created by scientists and engineers in the same lab that let the infections out (CREATION). They made it as human as possible so that it was able to move, think, and interact with people and the environment just like we do. It is capable of making its' own decisions and learning from mistakes (INTELLIGENCE) (HUMAN) (AUTO). It knows how to think for itself. Its objective is to collect blood samples from people and see how many are infected, how many are uninfected, and report back to us. For instance, it was able to grab your arm and take the sample on its' own. We only intervene when something goes wrong like some idiot pulling a wire."

"Sorry about that. While I finish up with this, can you tell me more? What did my sample say?"

"Well as you probably saw after word of the infection was released, the world fell into chaos. It looked like World War III out there with people fighting over resources and whatnot (WAR). Our creation has to follow simple laws just like us. It can't murder or steal, and it

Fig. 8 (continued)

work during class time is a waste, but I don't think they've thought about the ownership one experiences when reading their work aloud to a live audience. Students tell me all the time that their experience presenting at the roundtables helped them get over their anxieties about speaking in front of crowds, and as an audience, our ability to give quality feedback to the presenter is enhanced because we have heard the writer present their story in the tone and voice of their choice.

After our first series of round table workshops, students have had the chance to see how others have handled rules they struggled with, and they've witnessed first-hand how important it is to paint a picture for their audience and to engage them

has to be ethical in all situations (LAW 1-3). We discussed the use of machines like this for this purpose many times before, but some people were too stubborn or crude to listen or accept others' opinions on the issue (DEBATE) (BIGOT). We warned them that our creations would be irreplaceable if anything went wrong (WARNING 1). Then, when everything went down the toilet, we warned them again that our creation would be needed (WARNING 2). They scoffed at us but here we are. Our world was heading towards great things. Our society was making great leaps and bounds in technology—flying cars, nanotechnology, advanced medicine, etc. Then it happened, the only thing left of the technology from the old world is what's in that machine in front of you and the little we have in our lab here (HIST)."

"That's all great, but what about the results from my blood samples?"

Changing the subject, the voice says, "Have you ever seen the late September sunset over the lake? It really is a beautiful sight to see. The colors of red, orange, and yellow, still with a hint of blue sky. Striped clouds dissect the magnificent colors (BEAUTY). During the conversation I had refitted the wire and the creation stood up. Looking me eye to eye, it said, "*Vivir como te estabas muriendo.*" I guess I hadn't reconnected the large figure correctly, but it worked. Luckily I had taken Spanish in High School. I was able to figure out that it had said, "Live like you were dying" (TRANSLATION).

Solid advice, but why would it tell me this? Was something wrong? What did he know that I didn't? There is only one thing it could be…

I am infected, I think. *No I can't be. How could that have happened?* (NERV) While I don't think that I need to, I do anyway—I ask it if I am infected.

The sound of rain dumping down loudens with the roar of thunder (L/W 4&5). A solemn nod is all I need.

Fig. 8 (continued)

through humor or drama or intellect. They receive their feedback from me about a week after they present and then they have until the end of the semester to revise or rewrite their story. For the final draft, they are not required to follow the detailed list of genre rules. They must only present a polished version with a cover letter reflecting on their revision process.

We move on to the second round, repeating the entire process. This time, students know each other better and feel more comfortable writing for their audience. Their ability to work creatively within the constraints of an assignment is noticeably improved. By the time they submit their final portfolio, their stories generally feel more natural. This "natural" or spontaneous quality is emphasized as one of my primary goals with the assignment. I want students to be able to meet rigorous demands in their writing without having those demands take over their voice in the process. To me, this sort of fluidity is the mark of mastery that I look for in the final version.

Generating the Generative and Cross-Curricular Conversations in the University

In the United States, there are many different approaches to incorporating writing instruction at the college level. Some universities employ Writing Across the Curriculum (WAC) approaches by infusing writing into their curriculums at all levels and throughout general studies requirements and the major. As of now, Colorado State's *WAC Clearing House* webpage, http://wac.colostate.edu/programs/, lists dozens of universities with active and official WAC programs, but there are also many universities not listed here (like my own) that have seen their WAC programs evolve into Communication Across the Curriculum (CxC) programs which have similar though expanded aims, including oral communication and visual/digital literacies. Many universities employ a WAC philosophy without having a stand-alone WAC or CxC program, instead focusing on curricular requirements and Quality Enhancement Programs (QEPs). Other universities, like UC Davis, are combining the values of WAC with a Writing in the Disciplines (WID) approach to create "WAC/WID" programs. Universities like these require writing at all levels of undergraduate study but additionally offer discipline-specific writing instruction by drawing on the talents of writing specialists trained to work with students in particular majors. MIT's efforts to implement an "Innovation Across the Curriculum" program aimed at preparing their students for the writing tasks that will face them in their professions, extending the reach of writing and communication instruction to include oral communication, collaborative writing, and visual rhetoric is another interesting mutation of this hybrid WAC/WID phenomenon (Craig, Lerner, & Poe, 2008). This kind of connection between innovation and design thinking has been unpacked by thinkers such as Wylant in his 2008 article "Design Thinking and the Experience of Innovation" in which he argues that creative collaboration and "out of the box thinking" are paramount to quality performance in innovative fields (Wylant, 2008).

Such dedication to writing as it relates to innovation and professional preparedness is also in line with much of the research in the field of writing studies that has been recently published, establishing as it does, design thinking as central to the writing process. In "Design Thinking and The Wicked Problem of Teaching Writing" (Leverenz, 2014), mentioned earlier, Leverenz' argues in favor of presenting students with "wicked problems" in the writing classroom in order to stimulate cognitive development and foster creative thinking as well as problem solving and design thinking capabilities. Her reference to "wicked problems" such as those that appear as the sample assignments in this essay, stems from Churchman's "Wicked Problems" as perceived through the discipline of Management Science (Churchman, 1967). Long before Leverenz' 2015 article, Buchanan transposed the possibilities of "wicked problems" in the realm of Design Studies with "Wicked Problems in Design Thinking" (Buchanan, 1992). In 2001, Buchanan reframed the argument for *Rhetoric and Philosophy* in an essay entitled "Design and the New Rhetoric: Productive Arts in the Philosophy of Culture" where he unpacks the rhetoric of technology as one of design, stating that in designing products we should be aware that, as designers, we are making arguments for how others should live their lives (Buchanan, 2001). Marback's 2009 article, "Embracing Wicked Problems: The Turn to Design in Composition Studies," offers an explanation of design thinking and its benefits to teachers of writing who are faced with curricular shifts that emphasize digital and multimodal composition and require the blurring of form and content that occurs in the creation of such texts (Marback, 2009). A similar stance and extensive literature review of design thinking in relation to the field of writing studies is presented in Purdy's 2014 article, "What Can Design Thinking Offer Writing Studies?" where he tracks the occurrence of the term *design* as it appears in writing studies journals over time, attributing the rise of the term's occurrence to recent curricular efforts to include digital and multimodal writing (Purdy, 2014).

At my university, design thinking is taking hold as our curriculum moves to include studio time as part of the First-Year Writing requirement. We are making space for design in our writing classes as we expect more from the electronic portfolios and multimodal texts we ask our students to create. In addition to First-Year Writing, our students are required to take other writing-intensive courses in both their major and other general studies courses so, in the WAC manner, writing instruction is continually reinforced and recontextualized throughout their undergraduate experience. We have a QEP program at work in our general studies program that reemphasizes reflective writing and critical thinking skills across the various disciplines that comprise general studies. In terms of technical and scientific writing, many departments are members of our university's CxC program, though several departments manage their own WID efforts internally, so they vary greatly from department to department. In Mechanical Engineering, we have a writing component at the introductory level and then follow that requirement with a writing-intensive lab session at the sophomore, junior, and senior levels. At the sophomore level, grammar is dealt with quite directly while at the junior and senior levels, the focus is more on meeting the Specifications and Requirements set forth in our departmental *Formal Technical Reporting Guidelines* mentioned earlier here (The William States Lee College of Engineering, 2015).

In my current roles as lecturer in University Writing Programs and Technical Writing Advisor to the Department of Mechanical Engineering, I enjoy a kind of dual status, working both as part of a WID approach to writing instruction, one which is focused on giving students detailed and discipline-specific feedback on technical writing and particular engineering assignments, and as part of the WAC-inspired approach as it manifests in general studies requirements like First-Year Writing. The result, for me, has been a kind of stereoscopic perspective on teaching writing: on the one hand, I have to design writing classes for general studies courses aimed at cultivating design thinking and other habits of mind that I hope will prove useful when students write for other classes in the future—on the other hand, I provide upper-level engineering students with feedback that will help them do the kind of writing expected of engineers working in graduate programs or industry. Both roles have similar goals, but it was only after many long conversations with engineering faculty that I was able to see what the two objectives had in common and how the two approaches could enhance one another, how a design-based approach to teaching creative writing could help students becomes not only better storytellers, but better engineers. Since developing these teaching assignments, I have had many conversations with engineering faculty and with my colleagues in First-Year Writing about the usefulness of treating written texts as problems to be solved, and what I see is that, while there is a great deal of work to be done exposing the common ground we share, that common ground is most definitely there, infused into the creativity required in any discipline and in the writing process itself. Perhaps more conversations among such seemingly disparate worlds could help educators better attend to their students' needs.

References

Armand, H., & Weil, B. (2003). A new approach of innovative design: An introduction to C-K Theory. In *International Conference on Engineering Design.* ICED 03 Stockholm, August 19–21, 2003.

Buchanan, R. (1992). Wicked problems in design thinking. *Design Issues, 8*(2), 5–21.

Buchanan, R. (2001). Design and the new rhetoric: Productive arts in the philosophy of culture. *Philosophy and Rhetoric, 34*(3), 183–206.

Churchman, C. W. (1967). Wicked problems. *Management Science, 14*(4), B141–B142.

Craig, J., Lerner, N., & Poe, M. (2008). Innovation across the curriculum: Three case studies in teaching science and engineering communication. *IEEE Transactions on Professional Communication, 51*(3), 208–301.

Crickenberger, H. M. (2002). Project 1, stage 1: Course work for ARTS330—Advanced drawing with Roy Drasites at the University of South Carolina.

Crickenberger, H. M. (2015). Building a robot to do your homework: 'Crazy Concepts,' C-K theory, and design-based approaches to writing instruction. In *NC Symposium on the Teaching of Writing.* North Carolina State University, February 27–28, 2015.

Crickenberger, H. M. (2015). Pop culture as social critique (aliens): Requirements and specifications for UWRT1101—Introduction to writing and inquiry at the University of North Carolina at Charlotte.

Crickenberger, H. M. (2015). Pop culture as social critique (robots): Requirements and specifications for UWRT1101—Introduction to writing and inquiry at the University of North Carolina at Charlotte.
Faries, S. (2015). The fearless soldier (early draft).
Fritts, N. (2015). What is it? (early draft).
Garte, B. (2015). PT animal (early draft).
Leverenz, C. S. (2014). Design thinking and the wicked problem of teaching writing. *Computers and Composition, 33*, 1–12.
Marback, R. (2009). Embracing wicked problems: The turn to design in composition studies. *College Composition and Communication, 61*(2), W397–W419.
"Programs". *WAC Clearing House.* Colorado State University. Retrieved August 1, 2016, from http://wac.colostate.edu/programs/
Purdy, J. P. (2014). What can design thinking offer writing studies? *College Composition and Communication, 65*(4), 612–641.
The William States Lee College of Engineering. (2015). *Formal technical report guidelines.* UNC-Charlotte.
Wylant, B. (2008). Design thinking and the experience of innovation. *Design Issues, 24*(2), 3–14.

Writing as Knowing: Creative Knowing Through Multiple Messaging Modes in an Engineering Technical Communications Course

Jennifer L. Herman, Lynn Hall, Deborah Kuzawa, Leah Wahlin, and Mary Faure

In a recent commentary published in *The Chronicle of Higher Education*, Bordoloi and Winebrake (2015) argue that there are both structural and cultural barriers to integrating liberal arts courses into engineering curricula. Despite the advantages engineers gain from liberal arts-based courses, which are generally dedicated to fostering "design and creativity, teamwork and interdisciplinary thinking, and understanding the social, political, historical, and economic context of a project" (para. 2), Bordoloi and Winebrake contend that the changes to curricula and pride for the work done in traditional engineering disciplines prevent integration of liberal arts to the engineering programs.

As writing instructors housed within The Ohio State University's Department of Engineering Education in the College of Engineering, we support Bordoloi and Winebrake's argument in favor of integrating the liberal arts—including writing, communications, and rhetoric—into engineering education. Indeed, along with Winsor (1996a), we suggest that effective writing and communication are essential to the deepest learning of engineering concepts and are integral to the practice of engineering in professional contexts. The Accreditation Board for Engineering and Technology (ABET) seems to agree; included in its 2016–2017 criteria for accrediting engineering programs, "an ability to communicate effectively" is one of ten required student learning outcomes (ABET, 2015, para. 9). Further, *Forbes* has listed verbal communication, written communication, and other communication-related skills, such as the "ability to sell or influence others" as three of the top ten skills prospective employers sought in individuals who graduated in 2015 (Adams, 2014). Our mission in teaching engineering technical communications courses is to prepare students for the visual, written, and spoken forms of communication they

J.L. Herman (✉) • L. Hall • D. Kuzawa • L. Wahlin • M. Faure
Department of Engineering Education, The Ohio State University, Columbus, OH, USA
e-mail: herman.125@osu.edu; hall.1812@osu.edu; kuzawa.1@osu.edu; wahlin.1@osu.edu; faure.2@osu.edu

D. Bairaktarova, M. Eodice (eds.), *Creative Ways of Knowing in Engineering*,
DOI 10.1007/978-3-319-49352-7_4

will practice in their engineering coursework, in their job searches, and within the workplace.

We argue that learning and deploying effective technical communication does not simply mean learning professional writing genres and formatting conventions. Technical communications for anyone, but especially engineers, must be flexible and responsive to contexts in order to be effective. In other words, technical communication must be *rhetorical*. Communicators must understand that engineering and technical communications are "... always in relationship with a situation full of constraints, competing possibilities, audience factors, and purposes (often to influence behavior)" (Newcomb, 2012, p. 594). Effective engineering communication pedagogy should foster in students an understanding of a given communicative task's rhetorical situation: the interactions and relationships between the subject, context, purpose, and audience, all of which may be multiple and conflicting. For example, an engineer works to design and produce a new engine for a client; perhaps the engineer realizes the project will not be completed by the proposed deadline unless more resources are allocated. In this situation, the engineer must be able to analyze the context and communicate the bottom line to a variety of project stakeholders. She must be able to understand and explain how the project's timeline to completion and its quality might be affected if more resources are allocated, as well as if such allocations cannot be made. She also needs to be able to negotiate how to frame the bottom line for her manager (highlighting errors or setbacks in the design process) vs. how she should explain it for the client (highlighting how it will affect their financial bottom line or timeline for rolling out the new product), since each of those stakeholders—manager and client—has different concerns and interests related to the project. Similarly, providing the explanation to each of these stakeholders has a different purpose; for the manager, the bottom line helps in making a logistical decision about the best way to proceed and allocate funds, while for the client an explanation might work to preserve goodwill by allowing the client to reorganize the timeline for launching the new product. In this situation, contextual factors (a design setback, resources allocated, etc.) influence how the subject (the setback in resource allocation or timeline to completion) can be explained for various audiences (the manager, the client) for the engineer's purpose (explaining how a design setback can be fixed with more resources, or preserving the client's faith that the project will be done with high quality and efficiency).

Developing this type of rhetorical understanding can then help students more effectively and persuasively craft technical messages that are responsive to the contexts and needs at hand. The flexibility and contextuality of this liberal arts-based rhetorical approach to engineering technical communications can equip students with the intellectual skills to effectively assess and complete almost any communicative task. Rather than focusing on the recall of fixed knowledge, such as mathematics formulas or physics principles, rhetorical analysis requires a degree of creativity and an ability to assess dynamic situations and develop methods and modes of communicating tailored for each situation.

By teaching students an awareness of and strategies for tailoring their writing and communications for specific purposes, contexts, and audiences, we suggest they can more successfully engage with other engineering courses, concepts, and practices. For example, how might conceptualizing the *purpose* and *audience* for a chart or table help the student to better organize the information and use color and design to emphasize key interpretations? In what ways does imagining a supervisor's (*audience*) need to know urgent information upfront help a student to consider how to organize a verbal narrative with greater urgency, memorability, and emphasis? How can designing, building, and documenting a device for a particular *purpose* and *audience* help a team of students better understand the nuances of collaborating with others and presenting information? How can that same process help them understand the ways in which each mode of communicating—in writing the procedures, in sketching the design, in building the device—builds and layers together to create the overall, and infinitely more rich and rhetorical understanding of the final physical and written products themselves?

In this chapter, we demonstrate the value of conceptualizing writing and communicating about engineering and related concepts as integral to the process of *knowing* in engineering. Rather than simply teaching students to articulately record given facts or data, we suggest that teaching students to communicate using multiple messaging modes—the written, the oral, the visual, and the design or physical mode—also helps them understand the inherently *rhetorical* nature of engineering. One cannot write a document, design a device, or communicate a process without making choices about which data to include, what specifications to adhere to, which key ideas to emphasize—each of which is determined by the specific rhetorical situation of a piece and its mode of messaging.

We argue that the iterative, rhetorical processes students engage in as they build, compose, revise, and reassess their new knowledge are an integral process of learning, teaching, and knowing in engineering education and professional practice. In what follows, we describe three assignments that help students develop not only the tools and skills to communicate engineering and technical concepts, but also the understanding that all knowledge is rhetorical: contingent on purpose, audience, and the context in which it is communicated. Through this analysis we illustrate the integral nature of writing in the iterative process of consuming and producing knowledge that defines the teaching and learning of engineering technical communications concepts.

Chapter Structure and Content

In this chapter, we first establish our theoretical perspective through a review of the existing body of research on writing as knowing and the ways composing can interweave with the interpretation and production of visual images, data, and physical objects to enhance student learning. We follow this review with a presentation and analysis of three assignments from our engineering technical communications courses, each of which involves asking students to engage in rhetorical writing practices with the interpretation, design, and documentation of instructor- and student-directed end products.

Through our positioning of our engineering assignments in the framework of writing as knowing, we argue that instruction in all engineering disciplines should include activities that help students better "know engineering," developing an awareness of how the knowledge/texts they produce and deploy speak to and acknowledge their awareness of engineering contexts, conventions, and representative systems. Specifically, engineering assignments and instruction should model liberal arts instruction by including explicit discussion and understanding of the social function/purpose—how and why their formula, equation, product, improvement, concept, and so on—will be used by the audience addressed and how it meets their expectations (whether that be another engineer, a decision-maker, their grandmother, or someone else). We suggest that focusing on the rhetorical constraints will help students develop a meta-awareness of the task at hand that will help them better understand how to engage in shared knowledge-building practices that endemic to the writing scenarios in which they will find themselves as engineering professionals.

Review of the Literature

The following review explores scholarship on the ways in which writing and communicating in multiple modes helps communicators establish and understand what they know, as well as to better translate that knowledge for different audiences and contexts.

Writing as Knowing

Following scholars in technical communication and rhetoric and composition, our pedagogy is shaped by a view of writing as an epistemic process by which individuals not only express knowledge they already possess, but actively *create* and *build* their knowledge. It has been suggested that "…writing in the disciplines is founded on an integrative relationship between writing and knowing…both recovering knowledge and generating new knowledge" (McKeon, 1987; Miller, 2000; Carter, 2007, p. 386). In other words, disciplinary writing is *writing as knowing* and *knowledge*.

To be knowledge producers, Rogers and Walling (2011) assert, students have to be exposed to nonacademic genres and forms of writing, as well as educated on the contextual and interpersonal dimensions of knowledge production (p. 261). Teaching only generic conventions, such as those for memos, funding proposals, or white papers, falsely suggests that facility with these genres can be equated with successful writing and effective technical communication, and this encourages students to be passive users of knowledge rather than knowledge producers. Instead, our assignments focus on students establishing communicative situations in which genuine knowledge can be collaboratively built and shared among students. The key is helping students develop an understanding of how audience needs and communicative purpose influence what constitutes knowledge in a given writing situation. Through such scenarios, we aim to facilitate what Estreem (2015) describes as "…activity undertaken to bring new understandings. [Through these activities] writing…is not about crafting a sentence or perfecting a text but about mulling over a problem, thinking with others, and exploring new ideas or bringing disparate ideas together…" (Estreem, 2015, p. 19). Writing as knowing, then, involves learning to problem solve with others for particular, contingent rhetorical situations, a skill that students will be asked to engage again as engineers writing professionally.

In addition to approaching writing as a knowledge-producing activity, we also assume that what students already know influences what they are able to write. Effective writing pedagogy cannot involve simply releasing students into genuine engineering writing scenarios for which they are not yet prepared. In fact, Artemeva (2011) notes that students can become frustrated by technical communication coursework that invokes genuine engineering scenarios and concepts, but fails to meet students where they are in their academic progress. An example of this might be to ask a student to write a proposal designing a new application of signal processing before the student has taken a signal processing course (Artemeva, 2011). Teaching audience awareness, purpose, and context, but failing to acknowledge the level at which students have engaged their disciplinary domain knowledge will not be successful:

> To become a successful communicator in the knowledge society, individuals must be in possession of the professional knowledge that they need to communicate ('the what' of their profession) and the appropriate rhetorical strategies that allow them to deploy this information within their professional community or beyond it ('the how'). (Artemeva, 2011, p. 322)

Successful technical writing pedagogy requires imparting the *how* of effective technical communication but just as importantly, technical communication instructors must assess what other disciplinary knowledge students do or do not have to complete the task. Winsor (1996b) aligns with Artemeva, suggesting, "In order to write well, a novice has to learn to perceive and react to the world as expert writers do" (p. 168). For a novice, such as a student, to view the world as an expert, they need to understand the *how* of technical communication—the ways that a rhetorical situation impacts the *what* of disciplinary knowledge. As instructors we need to assess students' prior knowledge and provide activities that help students understand where their knowledge stops and how to gain the knowledge they need to be effective communicators for given topic and context. Therefore, our pedagogy approaches engineering writing in terms of access to prior knowledge and an ability to adapt and build upon that knowledge in effective and productive ways.

That access to prior knowledge has been explored by Yancey, Robertson, Paterson, and Taczak (2014), who divide student ability to access and utilize prior knowledge in writing into three possibilities: assemblage, remix, and new knowledge creation. With assemblage, students use "both knowledge and practice" and add "a limited number of new key concepts" to their knowledge base. Students who more successfully tap into prior knowledge and practice are able to remix that knowledge, along with newly acquired knowledge, to complete tasks. We contend that low stakes assignments such as freewriting, brainstorming, idea mapping, and small group discussions can be used to help students identify their current knowledge and how it relates to the task at hand. Additionally, some students successfully create new knowledge and practices for themselves as a result of encountering a "setback or critical incident" (pp. 103–104). The "setback" could be the students' recognition that they do not have all of the knowledge they need to successful complete the task. This can force them to assess what knowledge they lack and where and how they might develop it or how they might alter their approach, including the intended topic, purpose, audience, and context. We present students with a variety of scaffolded writing tasks that ask students to engage and build upon prior knowledge with the aim of teaching them how to access, adapt, and build upon knowledge—to move away from assemblage, into remixing and, ultimately, into an ability to critically reflect and respond to new challenges.

The ability to critically reflect and respond is a skill necessary across their engineering curriculum but is paramount to their ability to successfully write on the job. The effectiveness of writing curriculum for engineering students depends on its ability to teach students the skills necessary for newly professional students to enter into an engineering industry-specific discourse community. We hold that by providing students with a variety of writing contexts and scenarios in which they can acquire the "language" of engineering genres, as well as practice using the rhetorical strategies needed to successfully participate in and continue to develop skills within those genres, we are teaching them the necessary tools to adapt (and write) in action.

Representing as Knowing: Communicating in Multiple Messaging Modes

In addition to approaching writing as knowing, both writing *to know* and writing *as drawing from what we already know*, our pedagogy assumes that effective engineering communication is based not only on *writing* as traditionally understood (written words on a digital or paper page), but also *communication* in the many messaging modes we use today. As paraphrased by Ball and Charlton (2015), New London Group suggests all writing is composed of multiple modes—"linguistic, aural, visual, gestural, and spatial"—and at least one of these modes is used in "every piece of communication a human composes" (p. 42). We contend that engineering technical communications involves not only these elements of traditional writing, but also the other messaging modes—the visual, the oral, the aural, and the physical—used daily in the professional practice of engineers and scientists, including the creation of narratives, oral instructions, diagrams, graphs, 3D models, and video demonstrations, among others. Engineering students are already asked to communicate in these different modes in their disciplinary courses, but at the same time, disciplinary courses generally do not discuss these different modes from a rhetorical point of view: the how, why, and in what ways these different modes create meaning, create meaning differently or more effectively for different audiences and contexts.

Melding the knowledge from the liberal arts, such as rhetoric and rhetorical studies, with the disciplinary knowledge needed to understand and make sense of the data ultimately provides students with a larger toolbox for effective, contextual communication.

The visual, the oral, the aural, and the physical, as much as written text, represent and build what engineers *know*. As Rogers and Walling (2011) note, "Knowledge… is created by multiple genres (e.g., texts, diagrams, graphs, reports, personal correspondence). Constructing and communicating knowledge about any subject, whether it is art or science, requires the knowledge worker to read, transform, and manipulate symbols" (p. 269). Like Walling and Rogers, we contend that effective communication does not only involve writing clearly and concisely to achieve one's purpose and meet the needs of a given audience; our pedagogy also asks students to practice representing and building their knowledge through visual, oral/aural, and physical or "hands-on" messaging modes by synthesizing their knowledge and knowledge obtained through research on a given topic.

Significantly, our assignments rarely require students operate in only a single messaging mode. Rather, we follow educational scholars Kress, Jewitt, Ogborn, and Charalampos' (2001) theory that asking students to interweave multiple messaging modes results in more complex meaning-making (p. 11). According to their schema, the use of narratives, oral instructions, diagrams, graphs, 3D models, or video demonstrations in scientific and technical disciplines has developed according to their creators' and users' purposes and actions. Those purposes and actions rarely operate *only* verbally or *only* visually: a graph in a report is interpreted in writing, a demonstration is accompanied by a verbal description of principles being illustrated, experimental data are organized to show a trend in a table, and so on. As such, our assignments are designed to elicit products from students that reflect the study results of Kress et al. (2001): "Meaning is made in all modes separately, and at the same time, that meaning is an effect of all the modes acting jointly" (2001, p. 1). In some ways, this is a contextual, social understanding of meaning-making. Our pedagogy assumes that engineering professionals must understand how multiple messaging modes and points of view may be used simultaneously to convey their knowledge to a wide range of audiences on a range of topics.

Multiple Messaging Modes and Teaching Transferable Communication Skills

Winsor (1996a) notes that, until knowledge is shared—that is, until one's audience has accepted one's interpretation of the facts or data—what the individual "knows" cannot truly be defined as "knowledge" (p. 5). Effective technical communication involves not just the building of new knowledge, but also persuading someone else of the case at hand, according to Rogers and Walling (2011):

> In fact, the representation of knowledge, whether in a text or any other medium that uses a visual display, arises from the social engagement of actors who must display novel information in forms that are well recognized and meet the expectations of their peers. Because of

> the need to meet audience expectations, representations of knowledge must employ and adapt existing conventional representations for new tasks. (p. 261)

In line with this body of pedagogical scholarship, our rhetorically based assignments ask students to engage across a spectrum of composition contexts that demand attendance to these "conventional representations" in order to foster the development of a "mental schema" that will allow them to adapt to the discourse communities they find themselves in as professionals. Part of this process is reflection, from the brainstorming and inception to the completion of a project. Self-reflections and team reflections provide students with space to consider what they have done; why they have completed it in the way that they have; how they might complete it differently in the future or for a different audience or context; and how they might transfer the knowledge and skills from a particular context and assignment to others, both in our courses, in students' other courses, and in the workplace.

Further, Shipka (2005) asserts,

> When students are called upon to work within a multimodal task-based framework, questions associated with materiality and the delivery, reception, and circulation of texts, objects, and events are no longer viewed as separate from or incidental to the means and methods of production, but as integral parts of invention and production processes. (p. 301)

In other words, multimodal frameworks require students to better understand the various factors that impact the effectiveness and usefulness of a communication, product, process, or device, resulting in more rhetorically aware—and rhetorically effective—communications. This aligns with Beaufort's (2007) discussion of how discourse communities' knowledge are developed, maintained, and circulated (see Fig. 1). Communicators of particular discourse communities share knowledge about subject matter, rhetorical conventions, writing processes, and genre conventions, and all of these make up the knowledge and expectations of the discourse community (Beaufort, 2007, p. 19).

In the following sections, we explore three different engineering technical communications assignment used in our courses. Each assignment requires "…flexibility, creativity, critical thinking, and strong communication skills (particularly

Discourse Community Knowledge

↕↕

Subject Matter Knowledge
+
Rhetorical Knowledge
+
Writing Processes Knowledge
+
Genre Convention Knowledge

Fig. 1 Discourse community knowledge is developed through sharing knowledge of subject matter, rhetoric, writing processes, and genre conventions (Beaufort, 2007)

writing) [which] are at the core of liberal arts education and critical to success today and in the future" for engineers and nonengineers alike (Christ, 2012, para. 2); these assignments also require familiarity with disciplinary knowledge. The following assignments assume that for engineers to produce knowledge, they need to navigate and engage the situated and social nature of knowledge-making practices, such as telling their research stories, taking an idea and making it into a physical/material representation of those principles, and considering how the individual with whom they are communicating—their knowledge and interests and concerns—should and does influence the ways in which they can interpret, explain, and represent data that they generate in their research.

Making the Numbers Talk: Evaluating and Commenting on Data Sets

This section describes the Data-Driven Commentary Memo, an assignment that requires students to transform raw data into verbal arguments accompanied by graphic evidence. Although we are writing instructors housed in a college of engineering, many of us have academic degrees or extensive training in rhetoric and composition—the art of effective and persuasive communication and the study of how such communication can be taught. Rhetoricians tend to view all knowledge—even scientific knowledge—as contingent and dependent on context, audience, and even the terminology and ways of knowing endemic to disciplinary cultures that novices must learn to be considered experts.

This view of scientific knowledge has been well documented by both rhetoricians and sociologists of scientific knowledge including Latour and Woolgar (1979), Bazerman (1988), Gross (1990), and Knorr Cetina (1999). These scholars hold that what we come to accept as scientific data and facts are contingent on the tools we use to produce them and the theories and concepts through which we interpret them. When we argue about the meaning of a test result, or the explanation for a particular phenomenon, we reveal the contingency of scientific knowledge on our understanding of, and belief in, the theories that comprise our fields' disciplinary knowledge. Winsor (1996a) has suggested that generating engineering knowledge is similarly rhetorical, meaning that "…persuasive work is part of that activity and…knowledge is formed in interpersonal negotiations over interpretations of evidence rather than simply in the close individual examination of an unambiguous reality" (p. 5).

Yet in engineering, perhaps to a greater extent than in the sciences, the culture and courses in which students are immersed present the work of engineering as outside of the influences of argumentation, context, and contingent knowledge (Layton, 1971, as cited in Winsor, 1996a, p. 11). In fact, according to Winsor, "Engineering education is frequently directed almost solely toward teaching individuals to produce quantitative data and thus may leave young engineers even more unprepared than most college students to see knowledge as rhetorically constructed"

(p. 12). We suggest that engineering knowledge and expertise, like scientific knowledge, is indeed, rhetorical: acceptance of one's ideas and interpretations is based not only on one's technical facility, but on one's ability to persuade the listener or reader that one's interpretation or idea is the best solution. As such, technical writing and communication become a vital way of teaching engineers how to both *know what they know*, and, conversely, *how to transmit that knowledge to others*. As Winsor (1996a) asserts, until knowledge is shared—that is, one has effectively communicated what is known to another party who shares the same interpretation of the facts or data, what one person knows cannot count as "knowledge" (p. 5).

Given the culture of engineering and its orientation toward quantitative and verifiable data as constituting knowledge, how can engineering educators help students to complicate their understanding of knowledge and knowledge production? As instructors of writing, we constantly grapple with fostering in students an understanding of the contingency of knowledge, the rhetoricity of data and information, and the needs of audience; ultimately, our goal is to help students then deploy this knowledge in the various contexts in which they are asked to communicate.

One assignment through which we have had success in helping students develop these skills is called the Data-Driven Commentary assignment. The Data-Driven Commentary assignment is taught in an upper division Electrical and Computer Engineering course devoted to technical writing and presentations. In this exercise, students are provided with raw numerical data from two universities: The Ohio State University and University of Michigan. These reports provide information on the outcomes of graduates from their engineering programs. Data provided in both reports includes initial job salary data, percentage of students with job offers prior to graduation, percentage of students pursuing graduate degrees, and numbers of students who pursued an internship during their undergraduate careers. Students are directed to review, analyze, and interpret the data with a specific aim: they must use their interpretation of data in select criteria to compose a data-driven argumentative memo that combines written and visual interpretations. The goal of the memo is to provide data-driven evidence to persuade a high school audience to apply to the "best" program.

We ask the students to consider contextual factors that could influence how they could interpret the data and how their young audience might receive it. As a class, we discuss the following types of questions: What are the criteria that make one academic program "better" than the other? How does the descriptor, "better," change when we consider the needs and concerns of a high school senior vs., say, their parents, or prospective employers? What might be the intangible (i.e., non data-related) criteria available at each school, such as study abroad opportunities, diversity of the student body, extracurricular and internship opportunities, among others, and how those could those be weighted to influence how the data are interpreted or perceived? How has each school selected and framed the data they've provided, and how might that influence the way it should be interpreted?

These questions encourage students to consider criteria that individuals use to make decisions beyond those used to categorize the data. By contextualizing the information and its purpose and audience in this way, students are invited to act as a "spokesperson" for the data:

> …objects and data need a spokesperson who will stand between us and them to tell us what we see…different people can vie for the role of spokesperson for the same object or data because different people will produce different data or interpret the same data differently. (Winsor, 1996a, p. 2)

Indeed, students often choose criteria one would expect to see—starting salary, tuition costs, job placement success—as their primary criteria for decision-making, but they also select the intangibles, such as the geographic location of each school, the diversity of the student population, and the number of scholarships each school awards based on need or merit. Such selection demonstrates sensitivity to the emotional reasons, as well as the logical ones, that 18-year-olds choose one school over another. Students learn that data do not make meaning on their own, in a vacuum absent from contextual factors that also influence decision-making and what we think we "know" about something. This interplay of data interpretation, expert judgment, data representation, argumentation, and translation for non-experts helps students grasp that the numbers never speak for themselves. Rather, the "knowing" is rhetorical: it arises from the contextualized interpretation and representation of data and based on how the data is viewed and used by both the composer and the reader.

Rationalizing vs. Reasoning

The selection of criteria can sometimes reveal another tendency students have while interpreting this data: rationalizing with the data instead of reasoning with it. By rationalizing, we mean that students begin their interpretation of the data with a predetermined position; for example, they believe that "OSU's is a superior program to the University of Michigan's," and they then seek out the criteria and supporting data that bolster that claim, rather than beginning with the data and allowing the data to lead them to a conclusion. Rationalizing tends to arise more in this assignment than it might in others because the data we provide compares OSU and its largest rival, University of Michigan. The University of Michigan tends to perform better in the criteria that students, on first impulse, tend to assert should be the objective determiners for which program to select: in 2014, their graduates received more job offers and tended to be paid more in their first jobs (University of Michigan: Michigan Engineering & Engineering Career Resource Center, 2015). However, students at OSU tend to feel torn, because while the numbers in these criteria appear to "speak for themselves," students simply don't want the University of Michigan to come out on top. So students begin rationalizing: they cherry pick criteria and data in the reports that allow them to faithfully argue for their alma mater without pointing out the merits of its biggest competitor. For example, one team of students begins with strong reasoning, supported by the data. They argued that OSU offered a greater opportunity for local employment than the University of Michigan. This kind of interpretation is precisely the kind of thinking about data we want students to do: interpret the data in context, acknowledging that high school students in Ohio may be attracted to a program that will allow them to find employment after graduation near where they grew up. However, students who interpret data fairly and meaningfully can sometimes "cherry pick" data to support preconceived ideas they want to advance. Rather than continuing to compare data from both schools and providing context to explain their rationale for the significance of the context, they might report only OSU's data. Omitting the percentage of graduates from University of Michigan who receive job offers or pursue graduate education, a percentage higher than at OSU, is intentionally misleading, advancing the case for OSU without ethically representing the data.

When these situations occur during group work sessions in class, though, they actually become quite productive for emphasizing to students the difference between rationalizing and reasoning with data, as well as the ways in which data can be made to say all sorts of things that are not immediately self-evident. Though we steer students away from "cherry picking" or manipulating the data, we do like to call attention to how reframing the context or audience can change how the data could be interpreted. It is useful for students to understand and apply the ways the audience's values, such as living close to home after graduation, can lead to a positive, accurate interpretation of seemingly negative data. Similarly, for international students, the difference in cost between out-of-state tuition at OSU vs. University of Michigan matters differently than it does to a group of high school students from southwestern Ohio. Asking students to put themselves in the position of the audience in this way helps them to better understand how the data might mean different things for different people in differing contexts and situations.

Implications for Other Engineering Education Contexts

The success of the Data-Driven Commentary assignment hinges on its structure. To succeed, students must consider the audience for whom they are writing, and why and how they will select and frame the information provided in the memo. They must consider contextual and situational factors that would influence not only how they could interpret the data, but also how those interpretations might be received.

Engineering courses focused on design, among others devoted to teaching particular engineering concepts, methods, models, or equations could similarly benefit from asking students to attend to the rhetorical situation—the purpose, use, and user of the product, process, or information generated in course exercises. For example, in teaching engineering problem-solving, students may be asked to "create a visual representation to help understand the problem." However, without some concrete sense of the visual's user or her purpose (Is the user a client? The student herself? Another engineer? A government official?) the representation may be accurate, but it will never be as meaningful as it could be. By fostering students' ability to think about data and problem-solving rhetorically, engineering educators will not only help students develop a stronger sense of the contingency of knowledge, making them more comfortable in situations where uncertainty arises in problem-solving scenarios, but they will also expand and complicate what it means to "know" engineering concepts and "do" the work of engineering.

"Did You Know?" Lightning Talks: Telling Stories About Research

Structured as a short, informative presentation in the style of a "lightning talk," the assignment discussed in this section asks students to employ the creative and rhetorical conventions of storytelling and narrative structure to effectively communicate information from secondary research. In addition to its relevance to cultural trends like pecha kucha and TED talks, this style of presentation is rising in popularity at conferences, professional meetups, and in the workplace, where they serve as a way for employers to promote team building and knowledge transfer within their ranks, while giving employees an opportunity to enhance their presentation skills. The ability to present information concisely, in an engaging and well-structured manner, is vital for engineering professionals.

The assignment discussed here is intended to support knowledge transfer between group members of information built through research and to allow students to create knowledge through the process of recasting the information for multiple audiences. There are three important factors to consider regarding the context for this particular assignment in terms of the larger course activities and goals:

1. It is presented as a solo sub-assignment during the planning phase of an extensive collaborative group writing project; because of this, students are asked to use the presentation to communicate newfound knowledge to their teammates as well as to engage their other classmates (who do not *need* the knowledge for their own projects, but should find it interesting).
2. The presentation is developed in parallel with writing a more traditional informational report to communicate the results of the student's secondary research efforts in multiple modes.
3. During the term, students are also tasked with developing a formal group presentation that presents a public communication project or serves as a "pitch" for their funding proposal; the skills developed in the shorter, individual storytelling assignment aim to be applicable to those other, perhaps more familiar presentation contexts.

Here, the focus will be primarily on the presentation itself, but it is useful to acknowledge how these activities are situated within the larger context of the course as we consider how students learn to communicate effectively in written, visual, and oral modes, and transfer those skills to the various communicative situations they will encounter as professional engineers.

The assignment asks students to prepare and present a 2–3 min presentation or "talk" using five or six slides as visual support. In the spirit of pecha kucha format, in which the presentation is delivered alongside 20 slides that advance automatically every 20 s, these restrictions are emphasized and enforced. However, unlike traditional pecha kucha talks, students are allowed to advance their own slides. The time and the number of slides are arbitrary and can be adapted—the primary goal is to provide some formal parameters that inspire creativity, planning, and practice on the part of the students.

Within these parameters, the assignment gives students considerable freedom to design a brief, informative presentation in which they share information discovered while researching their contribution to a larger group project. Students are prompted to focus on some aspect of the topic they found particularly compelling or interesting that might complete the phrase "did you know ___ ?" and no specifications are given for amount or scope of information contained in the presentation. Students are not expected to include *all* of the information from their research, but to make rhetorical choices about what the information to include and emphasize in order to meet the assignment (and audience) parameters. By restricting the formal elements of the assignment and giving students license to design the content, they are being asked to think rhetorically and creatively about the needs of the audience and to "play" with the larger amount of information they have amassed.

This creative sense of play, exploration, and adaptation of the information is in itself a critical communication skill. The presentation becomes part of the iterative process of *creating* both the specific knowledge about the topic and the skill of modifying information based on the particular rhetorical situation, including the context and audience. The underlying intention here is to help students see that there is not one single way to deliver information. Both presentations and information are rhetorical: the information itself is dependent on specific contexts and what they want the audience to know, and there are multiple opportunities to convey the same information in different ways to different audiences for different purposes (such in a written report vs. a short oral presentation). This challenges the notion that "facts speak for themselves" and counters the notion that information can only be communicated in one way.

To provide some structure to the presentation, the idea of storytelling is integrated into the planning phase. We turn to the basic ideas of narrative structure—setup, conflict, and resolution—to help students think strategically about structuring information in a way that is thorough and engaging. Most students are familiar with these concepts in some way, as they are commonly taught in elementary and middle school language arts curriculum, but we take time to review them in class. To illustrate the discussion, *The Cat in the Hat* serves as an example, but any number of familiar children's books or movies would serve the same purpose. Using a familiar plot line, we review the basic elements of the plot and map them together as a class. Then, we shift focus to a TED Talk (Terry Moore's "Why is the 'x' unknown?" works well) and, first in small groups and then as a whole class, we discuss how the narrative structure can be seen in the structure of an informative presentation.

The intention behind this series of discussions and activities is to allow students to see the patterns that structure communication so that they can apply them to their own presentations. The goal is also to demonstrate that creativity is key to presenting information effectively and engagingly. The framework provided by the narrative structure is both familiar and flexible, in that it can be projected onto any number of situations.

- In the "setup," the presenter provides context or background information for their audience; from observation, we have found that this phase of a presentation is particularly difficult because it relies so heavily on assessing the audience's existing knowledge of and interest in a topic.
- In the "conflict" or "rising action," the presenter reveals the information, perhaps challenging the audience's assumptions—the key here is to be deliberate in conveying the information, not in a rush, but in a planned sequence of "reveals" that leads to the "climax," which is a reveal of the presentation's main point. The visual of the rising line of action serves as a reminder that the audience cannot absorb all of the information at once.
- Finally, in the "resolution," we discuss that an effective conclusion must also make clear why the information is valuable to the audience, how it might affect them or what the implications of this knowledge might be.

The following table is included in the assignment instructions and used to guide the discussion of planning the presentation content (Table 1):

Table 1 Material used to support planning of students' lightning talk

Setup	Introduces characters and basic situations, sets the scene	– Provide context for your audience, describe the basics
		– Describe assumptions/beliefs that you will challenge ("You might think that…")
		– Establish the way things are, starting with a common question or mystery ("Have you ever wondered why…")
Inciting incident→Conflict (rising action)	Something happens to change "the way things are" and introduce conflict to the environment and characters	– What is the new information? What did you discover?
		– Reveal the information deliberately, piece by piece
		– Build tension by asking questions, explaining how your understanding changed
Climax→Resolution	The action builds to a climax, then resolves; the world or characters might return to normal or be transformed	– Fully reveal the main idea (thesis)
		– Explain the implications of this new information (Why does it matter? Who does it affect?)
		– Reiterate your main ideas
		– Provide a conclusion for your audience

Since students have amassed a much broader range of knowledge about the topic than will fit in 2–3 min, they can approach the task of framing the presentation *with the mind-set of a subject matter expert*, selecting and recasting knowledge to meet the needs of their audience and the context. These processes are vital technical communication skills for engineers who, for example, might need to communicate with colleagues or a public audience outside their area of expertise for the sake of business planning or to convey safety information.

Finally, this assignment allows us to put the presentation slides in a secondary role and instead emphasize oral storytelling. Because students are only allowed five or six slides for visual support, they cannot simply put all of their talking points on the screen and read through them. They must carefully select a limited number of visuals that they think will enhance the presentation. We discuss the rhetoric of presentation elements: the impact of placement in the presentation, clarity of the visuals, and selecting visuals that add layers of meaning rather than duplicating what the speaker is saying.

Thinking about presentations as engaging stories rather than reading a report out loud is a skill that transfers to many other contexts. Considering how rapidly communication technologies are changing, this focus on the content and oral presentation can transfer from conference presentations to video-conferenced client meetings more easily. Using this assignment, we have seen students develop a much more audience-aware style by crafting informative yet attention-getting openings, sharing a wider variety of visuals, and adjusting the scope and pacing the delivery of the information. Using the familiar patterns of stories and the specific structure of the presentation helps shift the focus from the information itself to the rhetorical context and strategies for communication. Finally, thinking about communication as a rhetorical—strategic and context-based—process rather than a single end result encourages them to *consciously* draw on these skills in their professional lives.

Rube Goldberg: A Collaborative Approach to Building Multimodal Communication Skills

This section explores a team-based, hands-on public communications unit (aka The Rube Goldberg Project) in our engineering technical communications course, focusing on the multimodal process, communications, creativity, and failures that are required for successful projects. Fetzer (2003) contends that "the difference between an innovator and the average scientist [or engineer] is often the acceptance of failures and of risk-taking" (p. 598). The best engineers are creative—they identify novel ways to create or improve a process or machine, but novelty often grows from failures, from learning what *not* to do. Engineering, though, is also about logic and efficiency, finding the most straightforward and practical way to design or run a process or machine. In our course we ask engineering students to build Rube Goldberg machines, essentially asking them to do the opposite of standard engineering: identify an inefficient, complex, and fantastical way to complete a simple or straightforward task, concerned not with efficiency but with creativity and whimsy. Ultimately, the Rube Goldberg Project is not about the machine, but the team's process and communications about the process and machine. Our goal is for students to tap into their creative sides to develop fantastical machines, embrace failure's importance to success, and use various modes and technologies to communicate clearly with each other and a public audience about their process and experience.

Rube Goldberg machines are imaginative, comical, and overly complex machines that complete a (generally) mundane task. The machines are an excellent example of STEAM (science, technology, engineering, arts, and mathematics) education, or the bringing together of the arts and STEM disciplines. Because Rube Goldbergs often appear in secondary education STE(A)M classrooms, some engineering students are familiar with the machines and may have even built one themselves. However, not all students in engineering technical communications are STEM majors and not all STEM majors have been exposed to Rube Goldbergs. As a result, the process and product of Rube Goldbergs are new to a majority of students, so most of the teams are on equal footing in terms of knowledge and experience with the machines. Additionally, though students may have created a machine before, they may not have had to visualize their machine through storyboards and videos, and may not have had to communicate extensively about the machine, their inspiration, and their process.

The Rube Goldberg Unit requires student teams of 3–4 to develop a Goldberg machine, with particular focus on communicating about their process and end product. Teams must multimodally document both their successes and failures. The primary components of the assignment include:

- Design labs in and out of class to design and create a machine, including a storyboard.
- Video of the machine in action, including bloopers and failures.
- Website that displays the video, includes a vision statement, an About Us section, and most importantly, provides a narrative of the team's machine and process.
- In-class Show and Tell, a brief, casual, oral presentation of the machine using video and website.
- Reflective memo about the process, including an evaluation of teammates and the assignment as a whole.

Initial iterations of the Rube Goldberg Unit did not require students to focus on the team's challenges and failures when creating the video and written components of the project. However, discussions with individual teams and large class discussions revealed that the teams' design and process failures were the moments when students were required to rethink, retool, and adjust their processes and thinking about the project. In other words, the moments of failure required more creative thinking to identify the best process for completing the task and larger assignment goals. As a result, the assignment now requires students to document, in video and words, their failures. Matthew Syed (2015) argues in his article, "How creativity is helped by failure":

> The problem in the world today is that we only see the final product…What we don't see is the deeper story of how these innovations emerge. The tales we tell about creativity overlook this, too. We think of Archimedes shouting "eureka" or Newton being hit on the head

> by the apple and instantaneously inventing the theory of gravity. But these stories are pure fiction. They get the direction of creativity the wrong way around. Insight is the endpoint of a long term, iterative process, rather than the starting point. (para. 10)

Students are generally asked to provide instructors with their final, polished drafts and those drafts are then graded. However, this provides a false sense of the creation and communication process. As a whole, the Rube Goldberg unit works to peel back the shiny veneer of the final product to show the process, difficulties and all, by requiring documentation of failures and explicit reflection on the product *and* process. This impacts the rhetorical situation for the assignment. The topic is, of course, a Rube Goldburg machine, but the assignment's purpose is to display and analyze the moments of failure and success, not just the successes, which is commonly what students turn in to be graded. The purpose and context require a shift in understanding what a successful assignment or process might look like.

Reflecting on the moments when things did not go as planned forces students to consider the impact and importance of their failures for their project's (and team's) creativity and overall success. For example, a student team wanted to include the board game Mousetrap (itself a Rube Goldberg) at the start of their machine, The Ultimate Pong Shot. The team spent over 2 h trying to get their vision to work before they decided to revise their plan so that they could meet the Show and Tell deadline. The changes meant the machine was less complex in some ways; however, once the Show and Tell was over, the team returned to their original idea, determined to make it work. They viewed their blooper reels and discussed what went wrong before. In the end, the team was able to successfully incorporate the Mousetrap game by examining their failures and rearticulating what, exactly, they wanted the game to do for their larger machine. Without reflecting on their failures orally and textually, the students may not have figured out a way to make their vision a reality. The team's experience also demonstrates that successful engineering, regardless of the subfield, is based on failures and often creative responses to those failures.

Prior to the first design day in class, we discuss the assignment guidelines and rubrics as a class, and then we watch examples of Rube Goldbergs. We include a combination of both professionally produced Goldbergs (e.g., OK GO's "This Too Shall Pass" music video; scenes from *Goonies*, *Back to the Future*, and/or *PeeWee's Big Adventure*) and student-produced Goldbergs from previous course sections. Watching the elaborate professional examples and the smaller scale but still elaborate student examples provides students with a visual and aural sense of Rube Goldberg machines. We generally show student examples that provide visual evidence of the machine failures in order to emphasize the challenges of creating successful machines. In addition to watching examples, we have students explore previous teams' websites for their projects as this provides a sense of the rhetorical situation for the assignment, including textual components and expectations for the project overall. At the end of the unit, each student completes a meta-reflection on the process and team. Asking students to step back from what they've done, and explain in words why and how they made certain choices, what worked well and what did not, helps students to better understand their own processes. This is an example of writing as knowing: many times we do not quite understand what we have learned or where our knowledge gaps are until we verbalize or articulate them to others through talk or writing.

As noted earlier, the Rube Goldberg project is more focused on successful multimodal communication to a general public about the processes than on the machines themselves. Focusing on multimodal communication and the particular rhetorical context responds to the current cultural moment and societal expectations of what it means to be a literate citizen. As the New London Group (1996) contends, writing courses must attend to the various modes that students are already engaging with outside of the classroom. Similarly, Yancey (2004) argues that if writing courses do not require students to think critically and rhetorically about technologies and their roles in the writing and communication processes, students will continue to see technology as supplementary or external to the actual communication and persuasion processes. As a result, Yancey argues,

> … students in our classes [will] learn only to *fill* up those templates and *fill in* those electric boxes—which in their ability to invite intellectual work, are the moral equivalent of the dots on a multiple choice test. Students will not compose and create, making use of all the means of persuasion and the possible resources thereto; rather, they will complete someone else's software package; they will be the invention of that package. (p. 320)

Such a decontextualized approach to technology and communication—understanding technology as a neutral tool instead of a key player shaping the communication process—positions some modes, such as the visual or aural, as disconnected from traditional modes of communicating (i.e., alphanumeric writing). Though engineering is inherently multimodal, creative, and technology- and communication-heavy, engineering students may not think about or use those technologies in critical or rhetorical ways. The reflection component of the assignment especially provides the space for students to think about, for example, the rhetorical limitations of using a Wordpress or U.OSU.EDU template for creating the team's website, or the rhetorical limitations that the machine may inherently produce due to the action it completes (e.g., a machine whose purpose is to complete a pong shot will be of interest to a relatively small audience, primarily college students). The Rube Goldberg provides an opportunity for students to more formally engage in multimodal composing and embrace creativity and failure to enrich their learning processes.

At first glance, the Rube Goldberg assignment may seem a bit silly to students, especially when many engineering students are engaged in high-level, high-concept engineering design and construction. However, the assignment requires students to think and communicate rhetorically and critically about their processes and ways of communicating those processes to other, nontechnical audiences, and requires students to consider the role of multimodal compositions for knowledge-making and knowledge-application processes. In the end, the Goldberg project can help students recognize and develop other (rhetorical and creative) ways of knowing that will benefit their engineering education and careers.

Some Conclusions: Writing as Creative Knowing, Knowledge Transfer, and Engineering Education

Effective engineering technical communication is not just about writing effectively in professional genres and formats; rather, it is about deploying written texts, oral narratives, visual designs, and representations together in order to build messages constituting what students "know" about engineering concepts.

This discussion of successful multimodal assignments in technical communications courses suggests how asking students to compose in multiple modes allows them to creatively establish what they *know* about a subject, as well as how they might use those modes to most effectively communicate that knowledge to another person. In addition, presenting students with these communicative scenarios encourages students to respond to the needs of communicative purpose, audience, and situation, skills that can then be engaged and practiced in other learning and professional contexts as well. As Winsor (1996a) has suggested, the ability to tailor one's composing skills for a particular audience is dependent on continual activation of those skills in a variety of scenarios: "[Audience awareness] seemed to be a potential that was called forth more strongly by some situations than others. That is, it appeared to be triggered at least as much by social situations as by cognitive development" (p. 7). Winsor's observation suggests that the more student are asked to engage in these situated sociocognitive actions demanding their use of quality communication skills, the more they will develop (and value) those skills.

If practice of communicative skills in multiple contexts and situations can both help students more effectively know and communicate technical concepts, how can engineering education encourage the engagement of writing as knowing in contexts beyond the technical communications classroom? We argue that engineering programs should require technical communications and writing courses that approach the communication and knowledge of and about engineering as rhetorical. Further, we argue that engineering programs should foster the development of engineers' skill in responding to exigencies of purpose, audience, and context of communication in other technical and "major" disciplinary courses. We suggest that this be done in tandem with a dedicated technical communications program so that foundational rhetorical skills and abilities taught in such a program can be further practiced as instructors in disciplinary courses write assignments and frame activities that continue to reference and draw from those communication competencies.

For example, an architectural design instructor teaching students to draw schematic diagrams of space renovations could require students not only produce precise dimensional drawings of professional quality, but also a presentation of the drawing in a mock client meeting, wherein the student explains to the client how the renovated space addresses the problems with the original space, as well as how their redesign meets the agreed upon specifications of budget, materials, and so on. By asking students to consider how their drawing works together with their verbal explanation of its features and benefits, they must again practice skill with analyzing and addressing the needs of a particular audience (the client), purpose ("selling" the design), and context (the sales meeting). In this way, the disciplinary instructor can build the student's technical capabilities (schematic dimensional drawing, in this case) *and* further hone the student's ability to analyze and adapt to the unique demands of the rhetorical situation.

As they practice these communicative skills in tandem with developing their technical abilities, the value of these soft skills is reinforced and will clarify students' understanding of how these skills will be applied effectively in their daily professional lives. Christ (2012) argues, "Flexibility, creativity, critical thinking, and strong communication skills (particularly writing) are at the core of liberal arts education," (para. 2), so by bringing some of the liberal arts (i.e., rhetorical studies) into the engineering education classroom we can help our students hone the soft skills that make better engineers and that are desired by employers.

Although there may be structural barriers to integrating rhetorical skills into engineering courses and assignments beyond the dedicated technical communications courses, we assert that the modification of individual courses and assignments to call forth and engage students' negotiation of audience, context, and purpose would not be as onerous a task as the overhaul of an entire engineering program's curriculum. Engineering programs could foster collaboration between technical writing instructors and disciplinary instructors in order to modify assignments and lesson plans to incorporate discussion of audience, purpose, and context. Further, productive engineering education research could be done to assess the student learning outcomes of both technical and communicative skills in sections that adopted the changes.

Bordoloi and Winebrake (2015) have argued, "Engineering education needs to prepare students to design expansively and imaginatively—a capacity that liberal-arts education cultivates, along with critical thinking, clear communication, and productive collaboration" (para. 10). Our suggested approach to teaching and transferring rhetorical engineering technical communications knowledge to other engineering contexts is a step toward integrating liberal arts education into engineering programs and achieving those goals of more expansive and imaginative engineering design.

References

Accreditation Board for Engineering and Technology, Inc. (2015). *Criteria for accrediting engineering programs, 2016–2017: General criterion 3. Student outcomes*. Retrieved from http://www.abet.org/accreditation/accreditation-criteria/criteria-for-accrediting-engineering-programs-2016-2017/#outcomes

Adams, S. (2014, November 12). The 10 skills employers most want in 2015 graduates. *Forbes*. Retrieved from http://www.forbes.com/sites/susanadams/2014/11/12/the-10-skills-employers-most-want-in-2015-graduates/#3175028519f6

Artemeva, N. (2011). "An engrained part of my career": The formation of a knowledge worker in the dual space of engineering knowledge and rhetorical process. In N. Artemeva, M. Horne, A. Paré, D. Starke-Meyerring, & L. Yousoubova (Eds.), *Writing in knowledge societies* (pp. 321–350). Fort Collins, CO: The WAC Clearinghouse.

Ball, C., & Charlton, C. (2015). Concept 2: Writing speaks to situations through recognizable forms. 2.4. All writing is multimodal. In L. Adler-Kassner, & E. Wardle (Eds.), *Naming what we know* (p. 42). Boulder, CO: University Press of Colorado.
Bazerman, C. (1988). *Shaping written knowledge: The genre and activity of the experimental article in science*. Madison, WI: University of Wisconsin Press.
Beaufort, A. (2007). *College writing and beyond: A new framework for university writing instruction*. Logan, UT: Utah State University Press.
Bordoloi, L. M., & Winebrake, J. J. (2015, April 27). Bringing the liberal arts to engineering education. *Chronicle of Higher Education.* Retrieved from http://chronicle.com/article/Bringing-the-Liberal-Arts-to/229671/
Carter, M. (2007). Ways of knowing, doing, and writing in the disciplines. *College Composition and Communication, 58*(3), 385–418.
Christ, C. (2012). Myth: A liberal arts education is becoming irrelevant. *The Presidency.* Retrieved from http://www.acenet.edu/the-presidency/columns-and-features/Pages/Myth-A-Liberal-Arts-Education-Is-Becoming-Irrelevant.aspx
Estreem, H. (2015). 1.1: Writing is a knowledge-making activity. In L. Adler-Kassner & E. Wardle (Eds.), *Naming what we know: Threshold concepts of writing studies* (pp. 19–20). Boulder, CO: The University Press of Colorado.
Fetzer, J. (2003). The sky is not falling…accepting failure to create innovation in experimentation. *Analytical and Bioanalytical Chemistry, 375*, 597–598. doi:10.1007/s00216-003-1773-9.
Gross, A. (1990). *The rhetoric of science*. Cambridge, MA: Harvard University Press.
Knorr Cetina, K. (1999). *Epistemic cultures: How the sciences make knowledge*. Cambridge, MA: Harvard University Press.
Kress, G., Jewitt, C., Ogborn, J., & Charalampos, T. (2001). *Multimodal teaching and learning: The rhetorics of the science classroom*. London: Continuum.
Latour, B., & Woolgar, S. (1979). *Laboratory life: The construction of scientific facts*. Princeton, NJ: Princeton University Press.
Newcomb, M. (2012). Sustainability as a design principle for composition: Situational creativity as a habit of mind. *College Composition and Communication, 63*(4), 593–615.
Rogers, P. M., & Walling, O. (2011). Writing and knowledge making: Insights from an historical perspective. In N. Artemeva, M. Horne, A. Paré, D. Starke-Meyerring, & L. Yousoubova (Eds.), *Writing in knowledge societies* (pp. 259–274). Fort Collins, CO: The WAC Clearinghouse.
Shipka, J. (2005). A multimodal task-based framework for composing. *College Composition and Communication, 57*(2), 277–306.
Syed, M. (2015). Viewpoint: How creativity is helped by failure. *BBC News.* Retrieved from http://www.bbc.com/news/magazine-34775411
The New London Group. (1996). A pedagogy of multiliteracies: Designing social futures. *Harvard Educational Review, 66*(1), 60–92.
University of Michigan: Michigan Engineering; Engineering Career Resource Center. (2015). *Engineering career resource center (ECRC) annual report*. Retrieved from http://career.engin.umich.edu/wp-content/uploads/sites/30/2014/12/annualreport.pdf
Winsor, D. A. (1996a). *Writing like an engineer: A rhetorical education*. Mahwah, NJ: Lawrence Erlbaum.
Winsor, D. A. (1996b). Writing well as a form of social knowledge. In A. Hill Duin & C. Hansen (Eds.), *Nonacademic writing: Social theory and technology* (pp. 157–172). Mahwah, NJ: Lawrence Erlbaum.
Yancey, K. B. (2004). Made not only in words: Composition in a new key. *College Composition and Communication, 56*(2), 297–328.
Yancey, K. B., Robertson, L., Paterson, W., & Taczak, K. (2014). *Writing across contexts: Transfer, composition, and cultures of writing*. Logan: Utah State University Press.

The Engineering of a Writing Assignment: Optimizing the Research Paper in an Introductory Chemical Engineering Course in the United Arab Emirates

Lynne Ronesi

Introduction

This case study investigates professor and student perceptions surrounding an existing writing assignment—a research paper—in an introductory chemical engineering course at an American liberal arts university in the United Arab Emirates. This study grew out of earlier scholarship (Ronesi, 2011) that suggests that—in view of the academic writing challenges faced by the UAE student population from diverse educational, linguistic, and cultural backgrounds—discipline-based writing support should begin with lower division courses running concurrently with or immediately following first-year writing courses, with the support of an undergraduate writing fellow. While limited to one course, and indeed, only one assignment, this naturalistic study provides insight into the sociocultural and pedagogical variables that affect professor and student construal of this research paper and offers insight into adapting American university curriculum to local needs.

Leydens (2012) notes that "as an area of scholarly interest, engineering communication has begun to come into its own… [with] recent efforts hav[ing] greatly promoted the importance of engineering communication in multiple contexts" (p. 1). Indeed, over the past three decades, there has been a steadily increasing body of scholarship on writing in undergraduate engineering courses (Artemeva, 2005; Artemeva, Logie, & St. Martin, 1999; Dannels, Anson, Bullard, & Peretti, 2003; Herrington, 1985; McNair & Venters, 2012; Patton & Taylor, 2013; Peters & Arbor, 2014; Poe, Learner, & Craig, 2010; Winsor, 1996) and on communication in the engineering field (Pilotte, Bairaktarova, & Evangelou, 2013; Sageev & Romanowski, 2001). Poe et al. (2010) note "a national trend in engineering and science education toward achieving communication outcomes… driven by professional organizations

L. Ronesi (✉)
American University of Sharjah, Sharjah, United Arab Emirates
e-mail: lronesi@aus.edu

D. Bairaktarova, M. Eodice (eds.), *Creative Ways of Knowing in Engineering*,
DOI 10.1007/978-3-319-49352-7_5

that have recognized the importance of good communication skills for engineers" (p. 2), citing the support of the National Science Foundation and the American Association for the Advancement of Science 2008 Benchmarks for Science Literacy, as well as the Accreditation Board for Engineering and Technology (ABET). Yet, despite support from Writing in the Disciplines (WID) and Writing Across the Curriculum (WAC) programs and research, recommendations from national accrediting agencies, specifications from grant-funding organizations, and reports from the field that consistently highlight the importance of communication in engineering, introducing writing or communication modes into engineering courses has often met student, faculty, and institutional resistance. Poe et al. delineate the most common challenges to this integration—"disciplinary faculty not having a foundation in composition and communication practice and theory, institutions unable to create interdisciplinary partnerships between disciplinary faculty and communication specialists, and students not having flexibility in their schedules for communication instruction" (2010, p. 4).

These challenges, as highlighted in US-based scholarship, take on a bit of a different tenor when the writing in the engineering course takes place in English-medium universities in American universities outside of US borders—and indeed in this study—in contexts where students, and often faculty, negotiate vastly different sets of cultural assumptions, are used to different pedagogical modes, and lack confidence in their English writing skills. Given the strong presence of American universities and American branch campuses in the Gulf Arab States, and the fact that engineering programs are largely represented in their curriculums (Miller-Idriss & Hanauer, 2011), this situation bears study—in particular, naturalistic investigation—which, as Herrington (1985) posited in her own study of chemical engineering courses 30 years ago, allows us "by uncovering diversity, [to] see not only some of the problems that arise in specific classroom contexts, but also some of the possibilities—that is, ways teachers can create contexts conducive for using writing to achieve various objectives in a given course" (p. 333). This approach to scholarship is particularly relevant when a curriculum, with its attendant values and assumptions, is imported to a context where those principles are not necessarily shared in the same way.

In the Gulf region, research on engineering pedagogy in English-medium institutions is only just beginning to emerge. In the United Arab Emirates, Prescott, El-Sakran, Albasha, Aloul, and Al-Assaf (2011) report on an English Department technical writing course for engineers—a course developed in response to complaints from both employers and students who identified the level of intern and graduate communication skills as insufficient for the workplace, as well as to ABET recommendations for student engagement in group-based multidisciplinary engineering projects. Action research by Rudd and Telafici (2017) at a Texas A&M at Qatar (TAM-Q), an international branch campus (IBC) of Texas A&M, addresses the cultural and pedagogical complications of using American-authored textbooks and materials to teach first-year composition engineering majors—mostly students from Qatar, other Arab countries, and Pakistan and India. Another TAM-Q study by Hodges and Kent (2017) considers how WID or WAC programs can be localized or hybridized in the face of multilingual engineering faculty's frustration with the required writing in their courses and the regional students' comfort with more teacher-centered learning environments.

Seeking to build on and extend this developing body of scholarship, the semester-long, IRB-approved qualitative investigation presented in this chapter—informed by critical applied linguistics and WAC/WID scholarship—addresses the following research question: How do the professor, students, and writing fellow in this introductory chemical engineering course understand and situate the research assignment as a means of entry into the discipline of chemical engineering?

To undertake a naturalistic study of the writing assignment, I collected data using a combination of surveys, interviews, and observations. I interviewed the course professor, Dr. Zubeidi, three times over the course of the semester. In the first interview, I sought the professor's perception of the role of the writing assignment in the course, his understanding of the assignment's significance within the major and discipline, and his goals and expectations for student performance. In subsequent interviews, I probed those questions more deeply; I also took those opportunities to elicit Zubeidi's response to questions that arose during observations of the writing fellow and student conferences, and ultimately, to ascertain his impressions of the final drafts. At the beginning of the semester, I joined the class to observe Zubeidi's introduction of the assignment, and reviewed course documents such as the syllabus, the assignment posting, and a model paper from an earlier semester. I elicited student perceptions by administering an early- and a late-semester (postassignment) survey, observing student conferences with their writing fellow and interviewing the pair informally, and reviewing several submitted research papers by interviewed students. For interviews, I chose students whose survey-response comments or advice indicated a willingness to reflect on the assignment. The writing fellow assigned to the course, Marwa, shared her perspectives while I observed her conferences with the students on their drafts. All observations and interviews were recorded and transcribed. Consent was obtained by all participants, and all names mentioned herein are pseudonyms.

To support reader understanding of the sociocultural context of the study, I continue this chapter by providing historical background on the UAE, particularly those aspects leading to the UAE's current state as a diverse and globalized society which has adopted an English-medium western model for postsecondary education. I elucidate some of the struggle faced at the university to negotiate the space between its goals and sociocultural and pedagogical norms. Against that backdrop, I describe the development of the research assignment within the Introduction to Chemical Engineering course. Drawing on observations, surveys, and interviews, I present the concerns and perceptions of the professor, the students, and the writing fellow regarding the writing assignment; in particular, I highlight the interplay between sociocultural variables and curricular goals as students encounter the need to develop agency, acquire new knowledge, and start thinking like a chemical engineer. Finally, I discuss my findings in terms of adapting WID to local needs and, then, present changes made to the assignment in view of the findings.

The UAE and the University

The United Arab Emirates was, until 1971, a protectorate of England. With the advent of oil wealth, the visionary leadership of the UAE's first president, Shaikh Zayed, drew seven sheikhdoms, or emirates, into one federation. Due to the small, largely uneducated populace of the 1970s, the UAE's process of nation-building required and has continued to rely on an international workforce while it prepares its citizenry to take on the development of the country's infrastructure. A highly globalized environment resulted, and while Arabic is the official language of the UAE, English has become UAE's lingua franca to accommodate the communication needs of its global workforce. Yet, while English can appear to be foregrounded in the linguistic landscape of the UAE, in fact, it exists alongside of and interacts with the many languages of the diverse expatriate population who continue to maintain the cultural and linguistic expectations of their heritage country. In the UAE, this attention to heritage is well exemplified in the education sector. At the primary and secondary level, there is a plethora of educational models meeting the cultural, religious, language, and national curriculum requirements of the children growing up in the UAE. For national (Emirati) students, there are Arabic-medium government schools which offer some English instruction. For expatriate students, there are private schools which follow a multitude of national curriculums that accommodate the various expatriate communities. The level to which these schools teach English and provide writing instruction vary widely, but it is often the case that students graduate with fairly basic reading and writing skills in English (Shine, 2008, 2013).

As a tertiary institution in the UAE, the university accepts mostly resident expatriate and national students. A few students are international students from nearby countries. Most university students are native Arabic speakers from various Arab countries while many come from Pakistan and India. Like nearly all universities in the UAE, this university features an American curriculum with courses taught in English; however, this university is distinctive in its status as a private, coeducational, liberal arts American university. Accredited by Middle States Commission for over 10 years, this university has also received accreditation by the Accreditation Board for Engineering and Technology (ABET) for its Bachelor of Science degree programs in engineering. Indeed, the College of Engineering enrolls nearly half of all the university undergraduates. While engineering is a respectable and appreciated discipline all over the world, engineering has traditionally been understood in the Middle East and in the Subcontinent as very lucrative and prestigious; an engineer in the family is an economic and social boon that many families seek to ensure.

Following that, in this region, the notion of a liberal arts university is a somewhat uncomfortable fit. Ghabra and Arnold (2007) note, "[when it comes to postsecondary study in the Arab region,] there is an anti–humanities and social sciences culture that stems from the old thinking of strict employability within fields of study" (p. 12). This attitude exists in concert with a regional secondary-school approach whereby the teacher is the primary source of information. There is a deeply ingrained tradition of lecture-based, teacher-centered instruction which fosters learner dependency on teacher. "Knowledge, information and theories are often presented as indisputable facts, … feed[] on premises that do not accept dialogue or debate, [and do not foster the abilities] of associating, analysing, comparison, problem solving, [experimentation], or explaining facts" (Knowledge & Human Development Authority, 2010/2011, p. 40). (See also United Nations Development Programme, 2003, for discussion on MENA region pedagogy.) Dr. Zubeidi, the professor in this study, discusses his perception on the effect of this pedagogy on engineering instruction:

> [Students here] suffer because their high schools [are] based on spoon-feeding. And when you're spoon-fed, you do not realize you can find answers for yourself. Engineering design is open-ended. There is no correct answer. You have an infinite number of solutions. Now there are better solutions than others, but you have can have infinite number of ways to design a plant, for example. That open-endedness does not jive very well with the students here [because] they have learned there is one "correct" answer, which is a disaster because engineering is not about one correct answer. It's about how creative you can get, and this spoon-deeding destroys creativity. And why don't they read more? Because they are used to someone telling them what they need to know.

As Dr. Zubeidi suggests, a significant challenge to the fit of the liberal arts model in the region is student discomfort with reading. While teacher-focused primary and secondary classrooms may explain a marked student preference for oral instruction over other sources of knowledge, students' discomfort with reading, whether due to disinclination or novice-level reading skills in English, results in a reliance on the professor that runs counter to a liberal arts model. Miller and Pessoa (2017) address this quandary in their study with a similar student population at a university in Qatar. They report that students' novice reading skills and limited background knowledge so hinder students' understanding of content that they are forced to rely on professor's lecture in order to write about assigned texts.

Against this backdrop, the challenges of writing instruction at the university can be better understood. Like any American liberal arts institution, the university has a first-year composition (FYC) requirement. As noted earlier, the various secondary school models do not prepare the majority of students for studying at an English-medium, American-curriculum university, particularly in terms of academic writing. For most students in the region, writing in English or any language does not receive much emphasis at the secondary level, and very few have engaged in any source-based writing (Shine, 2008, 2013); as such, the FYC courses represent students very first exposure to the following skills: applying critical thinking to their reading and writing; structuring summaries, paragraphs, and essays; undertaking source-based argumentation; and complying with academic integrity requirements such as crediting authors. Given that these skills require time to develop, the university offers its first-year sequence in three, sometimes four courses (in the first semester, half of the students place into a noncredit developmental course). This length differs markedly from US-based universities in which FYC courses are contained in one or two semesters (Driscoll, 2011; Hall, 2006; Johnson & Krase, 2012; Melzer, 2014), and, to some degree, supports Ghabra and Arnold's observation that "American-style universities in the Arab world need to exert twice the effort as their Western counterparts to make a significant difference" (2007, p. vii).

That FYC both represents a steep learning curve and takes students well into their sophomore year has implications for student perception of their writing classes. Students frequently develop anxious or negative attitudes toward writing due to the sense of deficiency they develop in their FYC courses; this is particularly the case as many of the students, a number of whom graduated at the top of their classes, had been praised for their writing in high school. Moreover, students find that when they are asked to write assignments for other courses, the academic writing conventions and skills they learn in FYC are often not required. Bilikozen (2015), who studied six first-year regional students over the course of two semesters, points out, "in academic writing courses [the students] were expected to write formal essays, which usually required a clear argument, carefully organized supporting points, and integration of research following a standard documentation style, APA… [They would generally] produce several drafts, receive feedback from their professor, peers, the writing center, and finally revise their work accordingly" (p. 117). Yet, Bilikozen notes, when the students compare the writing assignment expectations of FYC to writing assignments in other discipline courses, they see that the assignment guidelines in discipline courses are less specific, drafting or periodic feedback is not required, and documentation expectations are far less rigorous; yet, student grades on these assignments are higher than their FYC grades. To many students, FYC seems an unnecessary torture.

This perception of writing as futile and torturous extends, as well, to some of the professor population. Bilikozen (2015) claims that the professors of the first-year students she studied attribute student writing challenges to both students and FYC curriculum. Most of the interviewed professors felt these difficulties are due to students' "lack of motivation to learn" compounded by little emphasis on reading and writing at students' homes, in their schools, or within their cultures. Moreover, some of those professors suggested the required FYC courses are "so basic that our students don't really get what we want them to get from them" (p. 161). While, in one case, that attitude resulted in the department's decision to address the problem by including some writing in all their courses, students' lack of writing proficiency can engender a reluctance by professors to require writing in their classes. On top of that, Ronesi (2011) reported faculty concern that some professors may not feel comfortable enough with their own English writing skills to assess student writing. Yet Ronesi's study also concluded that the steep learning curved faced by these students could be mitigated by writing assignments in lower division courses that run concurrently or dovetail with their FYC courses, especially with the engagement of a writing fellow. Sometimes referred to as writing consultants, associates, or mentors, writing fellows are undergraduate writing tutors who support student writers/writing in a particular course. While the uses and implementations of writing fellows vary from situation to situation, this model of course-based writing support encourages discipline-specific process writing based on student-writing fellow conferencing. (For more on the theory, practices, and concerns related to the writing fellow model, see Spigelman and Grobman's (2005) edited volume *On Location: Theory and Practice in Classroom-Based Writing Tutoring* and Hughes and Hall's (2008) special issue on writing fellows in *Across the Disciplines*.)

Introduction to Chemical Engineering: The Research Paper

One situation that corresponded to Ronesi's (2011) above-cited recommendation existed in semester two of a year-long Introduction to Chemical Engineering course. Several years ago, the originator of the assignment, Dr. Davenport—Dr. Zubeidi's predecessor—had planned to assign an informative paper in which students were to explain how an instrument integral to chemical engineering works and were requested to work with an undergraduate writing fellow. As Davenport and the writing fellow reviewed the assignment in their initial meeting, the writing fellow suggested the purely informative assignment be given an analytic component—further requiring students to suggest methods of improving the function of the chosen equipment. Davenport valued this suggestion and added more questions to the *Potential for Improvement* section, resulting in the following assignment:

Topic

Discuss an equipment used in Chemical Engineering Design in detail. These include but are not limited to compressors, turbines, heaters, evaporators, condensers, pumps, reactors, and separation columns.

Format

The paper should be approximately 8–10 pages long, double-spaced. It should consist of the following sections:

- *Introduction*—brief discussion of what the equipment or process is, what it does, and why it's important.
- *Description and Function*—how the equipment of choice is constructed, what materials are used, how large or small. Detail about what the equipment does, and how it works. What are common uses of the equipment?
- *Potential for Improvement*—what are the disadvantages to using this equipment? Do they use large amounts of energy? Do they produce large amounts of waste materials? Are there safety issues, such as explosive situations, toxic materials, etc.? Can these problems be improved upon? You may use information in the literature to consider improvements, but also think creatively about improving this equipment or process.
- *Conclusion*
- *References*

When Dr. Zubeidi began teaching the course, he continued with the writing assignment supported by writing fellows. Yet, he wondered how the assignment could be improved to broaden general knowledge of the field given the students' basic understanding of physics and engineering. "Would it be better," he mused during our first interview, "to look at a whole process, not only one piece of equipment? Or maybe talk about an industry? But that would become very general." He strongly felt—based on writing he had done as an engineer—that students should learn to write in a specific manner:

> [The current focus of the paper] is good because [students] actually have to sit down, they have to understand how [the equipment] works, and they have to articulate it, which is not easy. That's because it's very technical. And [the students] have to use technical terms and [explain how the formulas they already know apply to the process]. For example, when you're talking about a compressor, and how it works, and what the pressure does, and what the blades do. And so, that's what we're doing now.

To Dr. Zubeidi, the ability to describe how a piece of equipment works was very important, and, in light of one of the course objectives—"to become familiar with the various equipment and processes used in chemical engineering"—the informative component of the paper could be understood as an extended "writing to learn" exercise. Yet, Zubeidi also valued the ability to analyze the equipment critically, and accorded it a lot of weight in their grade.

> The most important part of the paper is the [analysis about improvement]. Because that [shows me] how they think. I want to see how they think. And it makes a big difference in their grade, too because the first part is all, you know, like, summarizing facts but [a lot of] of the grade is how to improve it—although it might be only one page. But this [part] will tell me if they're thinking properly and they're incorporating their own analysis into the paper, or if they're just copying facts.

One adjustment to the research paper process sought by Dr. Zubeidi in past semesters was to engage a writing fellow who was a chemical engineering major. Although he was always pleased with the nonmajor or generalist writing fellows he had been assigned, Zubeidi felt that a specialist writing fellow who had taken the course and understood the discipline of chemical engineering would prove more helpful to the students. I, who had just developed the writing fellows program when Zubeidi first took on the course, would have gladly accommodated his request for a writing fellow majoring in chemical engineering had one been available, but only on an experimental basis. Early in the development of the AUS writing fellow program, I had committed to a "generalist" writing fellow model and purposely did not assign writing fellows to courses in their discipline major. This decision was based on my concern that assigning a specialist fellow would lead to an emphasis on content correction rather than writing support. Also, as most of the courses requesting writing fellows were lower division courses, I had reasoned that generalist fellows could simply work closely with the professor to understand the discipline-related demands of the writing assignment. (For discussions related to generalist vs. specialist writing fellow and tutor approaches, see Dinitz & Harrington, 2013; Mullin et al., 2008; Severino & Traschel, 2008; Soven, 2001; in engineering, see Mackiewicz, 2004a, 2004b; Mackiewicz & Thompson, 2015.) As it so happened, Marwa, a senior studying chemical engineering, had been available to be assigned to Zubeidi's course in the semester prior to this study, and both Zubeidi and Marwa wanted to continue her support for another semester—the semester of my study.

Drawing largely on observation data supported by survey responses and interview data, the following section highlights how the stakeholders of this assignment—the professor, the students, and the writing fellow—mediate the tensions between social–cultural and institutional context, the challenges of American-style university writing, and the development of student agency veering toward engineering identity through their engagement in this basic discipline-based writing assignment. From Dr. Zubeidi's announcement of the assignment to class, to the writing fellow and student conferences, to student reflections of their struggles and reactions to the assignment, the data elucidate the negotiations of stakeholders in their attempt to connect seemingly disparate identities and territories to develop student agency and a chemical engineering identity.

The Stakeholders Speak

Dr. Zubeidi: Goals and Approach

A well-liked professor with a reputation for high standards as well as a scholar of international repute, Dr. Zubeidi was schooled in French and Arabic in the Middle East, and then completed his Bachelor's, Master's, and Ph.D. degrees in the US. After three degrees and some years of experience as a professor in the US, Zubeidi's speech is largely American in tone, with a light Arabic cadence. Zubeidi says of English: "It's not my native tongue, but it became my native tongue over the years." He attributes the beginning of that process—at least the writing aspect—to his student days in the US when he made ample use of the university writing center, an experience he recalls with fondness.

While Dr. Zubeidi inherited rather than created the course's writing assignment, he values the writing assignment and sees himself as responsible for introducing students to written discourse in chemical engineering. This attitude has prepared him to support the assignment against some challenges. At the department level, he has had to defend the research paper against faculty members' suggestion to replace it for more chapter coverage. In this capacity, Zubeidi invoked the assignment's role in responding to ABET criteria. The assignment, Zubeidi pointed out, supports the departmental plan to respond to the ABET criteria of "ability to communicate effectively" and "lifelong learning," explaining that the assignment fostered "student inquisitiveness—the students want to learn more about what they do [and] about the equipment." While the significance of meeting ABET criteria to the Chemical Engineering department, the College of Engineering, and the university itself made his defense unassailable, Zubeidi understands that the prospect of assigning and assessing written assignments makes many of his colleagues uncomfortable. "Honestly, whoever teaches this course has to be—and I consider myself—a native.... You have to have a native if you want to give writing the proper weight."

It is acknowledged that terms like "native" and "nonnative" speaker are inadequate to describe multilingual individuals in linguistically complex settings like the

UAE (see Geller, 2011; Nebel, 2017), but Zubeidi's statement suggests he feels uniquely situated to introduce the students to discipline-specific writing, sensing that professors with his level of comfort with, proficiency in, and enthusiasm for writing in English could best influence the chemical engineering students to prioritize writing as they prepare for their careers. Demonstrating this level of comfort and appreciation for writing is necessary to meet the resistance of students who feel "engineering is math and programming"—an attitude Zubeidi feels he must counter, to "make sure that the students know that engineering is going to be writing. Because, if they cannot communicate to others, they're useless." To launch his campaign, Zubeidi devoted an entire class during week two of the semester to introduce the research paper. In this class, Zubeidi led the students through the structure and requirements of the assignment, referring to supporting documents posted on the course-support software—the assignment explanation, the rubric, and a model paper. He also introduced the class' writing fellow, Marwa. Both the writing fellow and I attended this class.

As I observed this session, it was clear that Zubeidi actively sought to establish within the class a chemical engineering community unique to the realities of the students and the region, and to introduce, within that framework, an ethos of writing. In establishing the significance of writing to the profession, Zubeidi drew upon his understanding of the sociocultural and pedagogical context of the students to critically engage students in making connections between their currently disparate communities as novice writers and scholars and the imagined one of chemical engineers. As such, his community building was multidimensional, seeking common ground by establishing personal, cultural, rhetorical, and professional connections within the group, in part, by emphasizing his Arab heritage, his need in the past to adopt English, and the current challenges he—their professor and author of many scholarly works—still faces when writing.

Assignment Introduction

Dr. Zubeidi prefaced his assignment introduction by acknowledging that even though "we chemical engineers"—a communal reference he used a lot during his introduction—generally design whole processes, the students would narrow their focus in their research paper by investigating a piece of equipment used in chemical engineering. "This equipment is your whole life. We'll be learning about every single piece, all the mathematics, all the physics." Zubeidi detailed his expectations for the different sections by encouraging students to ground their choices in the careers they imagined for themselves, and further elaborated by "talking through" how students might approach writing the different sections: "Let's say you choose reactors. [You can state that] I chose to write about reactors because I am interested in such-and-such industry." Zubeidi then addressed *Part Two: Description and Function*—seeking student input to support the explanation. "How does a pump work?" he prompted. Most students seemed a bit surprised by this question, but one or two answered "By pushing? Force?" "Okay! Force divided by area equals what?"

Zubeidi asked. "Pressure," answered a few. Starting with the concept of pressure, Zubeidi scaffolded the class into a brainstorming session of pump operation. This exchange culminated in a discussion of the loss of momentum that occurs when fluid flows through pipes. Zubeidi reminded students that notion of momentum was treated by "the first law of motion from physics," demonstrating how to integrate physics laws into the description.

To introduce the idea of a disadvantage or inefficiency associated with pumps—the types of equipment shortcomings that would form the basis of *Part Three: Potential for Improvement*, Dr. Zubeidi asked students what would cause water to slow down in pipes, prompting a discussion on friction. He explained to the class:

> This is the part that I love the most because that's the part where you synthesize. And I want to hear about it. I want to hear how you can improve this particular piece of equipment.

Dr. Zubeidi demonstrated zeal for inventive ways of responding to challenging inefficiencies through providing a variety of interesting examples. Above all, he highlighted how these problems translate into wasted energy, which translated into higher costs. "Remember what I told you last week: We chemical engineers put the dollar sign in chemistry." Understanding that it might prove difficult for students at this level to come up with their own solutions for improving equipment, Zubeidi suggested that students address this part by searching the different literature on their equipment choice, and analyze and synthesize the responses to various shortcomings. Yet, he strongly urged students to venture their own responses once they had become conversant with the parts and process of their chosen equipment.

Secondary concerns then received attention: documentation, credible sources, plagiarism, assignment length, and introduction of the fellow. Zubeidi explained that using IEEE style for citations was preferred so students could gain familiarity with engineering documentation, but he indicated a willingness to accept APA, which is the documentation style students are taught in their FYC courses. Zubeidi then highlighted the need for credible sources—textbooks, journal articles—"Wikipedia will not be accepted as a source" and addressed plagiarism. While emphatic about his lack of tolerance for plagiarism, Zubeidi tempered his message with humor: "I've read many student papers and I know what you guys do: in one section of the paper, you sound like a student and in the next section, you sound like Einstein," which elicited some laughs. He also warned them that the discovery of plagiarism would result in a report to the Dean—"and I don't want you guys to have to go through that."

It was at this point that Dr. Zubeidi took the opportunity to introduce Marwa, the Writing Fellow, who was sitting in this class: "Marwa is graduating soon and has completed her Senior Design project, and she knows very well how to write in an engineering manner. She's a chemical engineer and will be helpful with both content and writing."

"Yes, I will let you guys know if you screw up [explaining] the thermodynamics," Marwa joked.

"You're lucky to have her help, because writing can be hard," Zubeidi added. "When it comes to scientific articles, English is my first language, but I too, sometimes struggle. Sometimes I spend a half hour on a sentence." This segued into a "plug" for time management: "That's why I am introducing the paper to you the first week. Start it now. Because the first weeks of the semester in engineering courses are pretty quiet. By the time the first draft is due you will be having midterms."

Fielding several student questions clarifying page length, Zubeidi engaged in a bit of contrastive rhetoric humor: "Don't be verbose. 'Verbose' means to write too much." He grinned, "I know there are a lot of Arabic speakers in the class. In Arabic, it is good to write more. If you can use two words to describe something, that's better than [using] one. And three is even better." The class laughed. "But English is not like that. And I want you to be concise, to write in a chemical engineering manner. Marwa will help you with that."

Finally, Zubeidi introduced the rubric and he used the opportunity to counter student perceptions that writing evaluation is subjective. "It's not subjective." He argued. "It's mathematical.... Sometimes when I advise first-year students and I note they have done badly in writing courses, they often say the professor does not like them. This is not the case." Zubeidi led the students through the rubric calculations, demonstrating painstakingly how their grade would be reached. "As you will see throughout the course, my goal is only to help you. You are like my little siblings and I want to help you out."

In this presentation, it was clear Zubeidi valued this writing assignment. He knew his audience, and his personal and humorous references to the realities surrounding their struggles to adapt to an academic writing ethos and to their eagerness to enter the chemical engineering world created an upbeat classroom atmosphere of optimism regarding the assignment.

The Students

The 28 students in Dr. Zubeidi's class were primarily sophomores (nearly 80 %), with one second-semester first-year student and a few juniors. All students had taken required introductory courses in chemistry, math, and physics, and the first section of the course. In terms of their FYC courses, about 40 % had completed FYC 3, the final course in the sequence in which students undertake an argumentative research paper in a supported process; nearly 60 % had completed FYC 2, the penultimate course in the sequence, which covers data search, source-based writing, and synthesis, and many of those students were currently enrolled in FYC 3. Only one student in the class, the lone freshman, was taking FYC 102 concurrently with Dr. Zubeidi's course. While taking the two courses concurrently was technically allowed, this rarely happened. Also, a bit out of sequence, a few of the juniors were concurrently taking English for Engineers, a class taught in the English department requiring FYC 3 as a prerequisite and junior standing. English for Engineers required engineering students from different engineering majors to work on projects.

Student responses to the first survey I administered (see Appendix 1) indicated that all respondents (22 out of 28, 78 % responded to both surveys) envisioned writing in their future capacity as chemical engineers; all expected to write some kind of reports (lab reports, progress reports). Sixty-seven percent anticipated writing proposals; 28 % expected to engage in research writing. After that, students mentioned a variety of writing genres: presentations, summaries, briefs, memos, analyses, agenda, meeting preparation, emails, and collaborative writing. It should be noted that students responded to the first survey during the week that followed the introductory class session, so it is impossible to ascertain whether students had these notions prior to the assignment introduction.

Nearly all of the class—90 % of the respondents—reported feeling at least somewhat confident about their ability to write in their field, even as most had very clear ideas about the steps or practices that would bolster their confidence. While nearly every response underscored "more writing assignments" as a necessary step, two secondary response trends equal in weight were "reading more discipline-based writing" and "knowledge of the field." Some fairly characteristic responses follow:

- Writing more research-based essays that would fit the criteria that are expected of chemical engineers.
- More in-depth knowledge of the research process and how to write research papers. Also by exposing myself to already published papers in chemical engineering.
- Reading more scientific literature.
- More writing in discipline courses. I've done four major courses and three of them required writing either a lab report or a research paper. At the end of that course, the paper I wrote is the most acknowledged thing I get out with.
- Being able to completely understand my limits when it comes to plagiarism and to master citation and referencing.

One technically faulty question on the survey was the one that queried students about the similarities and differences between writing for writing courses as opposed to engineering courses. Understandably, most students could not respond to this question as the "Introduction to Chemical Engineering" sequence constituted the first among their major courses. However, the few juniors in the class had already taken some core discipline courses during the summer and were able to offer commentary:

- In writing courses, professors focus on improving the students and noticing their performance throughout the semester. They help students to be on the right track. Discipline professors rely on the student's background on writing and expects the student to be already familiar with research papers and reports. Discipline professors do not spend a lot of time explaining what they want.
- In [writing classes we were taught] the importance of analyzing the material we had to research on and improving on the available data to form new and innovative thought processes. Although it might seem like an even more of an important concept in engineering, most papers/reports I have written as of yet, have not specifically asked for this criteria. Having said that, Dr. Zubeidi did emphasize that he is looking for that criteria and I am looking forward to the deeper study of chemical engineering materials.

The second, or postassignment, survey data (see Appendix 2) indicate students generally enjoyed the challenge of the research paper and found it meaningful: 90 % perceived the assignment increased their understanding of chemical engineering as their profession; 65 % felt the assignment helped to prepare them for the type of writing they will do as a chemical engineer; 70 % claimed they had put a lot of effort into the assignment; and a large number, 85 %, agreed that Dr. Zubeidi had expected high-quality writing in their assignment.

Students also acknowledged the contribution of Marwa—the writing fellow—to the writing assignment. In their postassignment surveys, 60 % of students agreed or strongly agreed that their discussions with Marwa helped them meet the requirements of the paper. In an open-ended question which asked them to list the additions/subtractions/changes they had made in their papers based on their two conferences with Marwa, students described Marwa as helpful in organizing the paper and in turning their focus to the applications, advantages and disadvantages, and to the challenging section on improvements.

Both the first and second survey data were further elucidated through the observation of draft one and two conferences with Marwa and through postassignment student interviews. These aspects of the study are highlighted later and bear witness to negotiations made as students start moving between zones—between agency avoider and agency seeker, between unknowing and newly knowing, between novice writer and proficient writer, and between student and chemical engineer.

The Paper Process

During this study, Marwa, the writing fellow, was in her last semester as an undergraduate chemical engineering major. Marwa spent a lot of her childhood in the US before her parents, of Arab origin, moved their family to the UAE when Marwa was in high school. A very proud and proficient speaker of Arabic and English, Marwa moved easily between the two languages although, like Dr. Zubeidi, on an academic level she was more comfortable in English. Dr. Zubeidi was clearly pleased with Marwa, highlighting her familiarity with the university's chemical engineering curriculum and her discipline-based knowledge: "She knows exactly what they are going through and the mistakes that they [will make]."

From her vantage point as an advanced student in the major, Marwa was able to put the role of the assignment into perspective in terms of its place in student learning and shared with students how and why the knowledge generated by the assignment would be useful when they began to work with some of the equipment. This is a point she would make to students at relevant junctures in their conferences or in the short chats that often followed a session. Marwa highlights her perception of the assignment's usefulness to and difficulty for the students.

> This assignment is very useful in that students are being introduced to equipment that is going to be really important to them in the future. So, rather than learning about something for the first time as a junior, they already have familiarity with it as a sophomore. That's also

> what makes the paper difficult for them, but it's exactly what makes it valuable for their junior year. It's difficult that they have to break the ice on their own—I definitely would prefer to break the ice with challenging topics like this in class with the professor—but at the same time, what you learn on your own is probably what stays with you.

Having been in their shoes 2 years earlier, Marwa knew what the students understood at this level, and this supported student learning in different ways. One important way in which Marwa needed to intervene concerned the location and incorporation of sources. Student had a very difficult time finding information they could understand. They found most literature concerning chemical engineering equipment to be extremely technical. "Even in the section of the paper where they simply have to explain the process, it might be that they read about the equipment and have a hard time understanding it and paraphrasing it. Journal articles don't really work for that part of the assignment. Most of them are 'over my head,' too," Marwa noted.

One source that many students found more accessible and reader friendly than the others was Wikipedia, which Dr. Zubeidi had warned them away from both verbally and in writing on his assignment sheet. Similarly, Marwa warned students as well during the conferences, but because of her experience with her own research, she had alternative suggestions for them, directing them to a group of engineering-specific reference books in the library that included encyclopedias and volumes that treated equipment.

While Marwa did not provide the students content, she was able to use the information already in their papers as a springboard to including more relevant material. For example, when Alia, who was working on centrifugal pumps, asked if she should include more equations, Marwa was clear on why and how:

> Yes, you should do that—at the very least, Bernoulli's equation. One of your pics here demonstrated centrifugal force, right? So you could just discuss centrifugal velocity and centrifugal force. You can demonstrate them with equations, and then add Bernoulli's. So yeah, equations help.

Relatedly, Marwa's familiarity with the equipment provided students with a critical reader of the drafts and allowed Marwa to question students in specific ways to elicit the information Dr. Zubeidi expected. During a first draft conference, Marwa could see that Mai's description about the process of a displacement pump lacked focus and was able to query Mai specifically to prompt for greater clarity: "So, how can we ensure uniform flow of fluids? And why are you bringing temperature into this description when a pump moves fluids at any temperature?"

Often, informal chats developed after draft discussions, and Marwa frequently fielded broader but related questions about upcoming upper level courses. At the end of a rather animated chat, Khalil expressed appreciation for the chance to "pick Marwa's brains" complaining "there was no one that [he could talk about his concerns in that regard]." These exchanges had no direct bearing on Dr. Zubeidi's written assignment, yet they served to enforce the idea of writing's potential for engendering a sense of community among the novice chemical engineers.

After a session during the second round of draft conferences, an interesting debate between Marwa and a student named Hashem arose concerning the use of the engineering documentation style, IEEE. When Marwa expressed concern about whether Hashem had correctly cited an encyclopedia reference in APA, Hashem protested the use of APA in the assignment, pointing out that IEEE was covered only in junior year, in the English for Engineering class, when by sophomore year students were already writing research papers and lab reports. While Dr. Zubeidi had offered the students the option of using APA or IEEE—although encouraging the use of IEEE—all of the students had chosen APA. Marwa and Hashem surmised that students did so because they did not understand the choice they were given: many students may not even have known that IEEE was a documentation style for engineering. Marwa admitted that she, herself, as a senior, had only used the engineering documentation IEEE in the English for Engineering course. She had not even used APA for her senior design project—"we were just so used to APA and professors didn't care either way." She points out "even Dr. Zubeidi doesn't have a preference now; he tells them they can use either." Initially, Marwa's response relegated documentation issues to an English concern than a discipline concern.

> I think it's important that we know it [but] honestly, there are more important things to worry about in engineering classes than citations. Yeah, in English for Engineers, obviously you should—that's what we're doing—it's an English course. But if you're writing a lab report, it's the least important, well, not that it's the *least* important, but it's more important you get the calculations and the concepts right, and you show your work.

Hashem argued the use of IEEE was more professional and termed the lack of its emphasis by the engineering faculty to be "a discrepancy." Even so, while Hashem knew about IEEE from an older brother, a chemical engineering major who had attended another university, and acknowledged Dr. Zubeidi's wish for the class to use IEEE, Hashem's resistance to adopting IEEE for this assignment—and hence his complaint—centered around his feeling that the students should have been taught the documentation by Dr. Zubeidi during class or by his lab assistant during a recitation.

The revelations made by students during these discussions on IEEE shed light on two issues. One, this conversation reflected entrenched territorial attitudes about the place of documentation in the chemical engineering curriculum. While it was unclear why the chemical engineering professors were relaxed about the document styles students used in their assignments, to Hashem, the use of IEEE conveyed an important link to their profession. Marwa's initial response, even as the assignment's writing fellow, demonstrates more territorial thinking between disciplines (Annous, Nicolas, & Townsend, 2017) suggesting that citation is a "writing issue" unconnected to content.

That both Marwa and Hashem concluded it was Dr. Zubeidi's duty to teach and require IEEE highlights a dependency on professors that many students, even strong ones, demonstrate. This dependency was highlighted many times during the conferences where students deflected responsibility for making learning strides onto the professor. Certainly, it was reasonable of Dr. Zubeidi to expect a motivated and rather savvy student like Hashem to pick up the IEEE documentation format on his own; that expectation was even more valid for the several junior students in his class who had taken English for Engineering, as they had reviewed the IEEE documentation and were required to use it in their course project. One such junior, Mai, explained opting out of Dr. Zubeidi's invitation to use IEEE for this assignment because "I have used APA for all of my studies." Indeed, this attitude was demonstrated in different aspects of the assignment. Kenza, whose abstract for her research lacked the appropriate information, excused herself with the following: "I don't really know how to do an abstract. I only learned about it last week in English for Engineering class when the professor was speaking to us about it. This was after already submitting my paper to Dr. Zubeidi." Even though Kenza expressed in her survey that she enjoyed writing and wrote on the value of learning on one's own by researching, it did not occur to her to investigate how to write an abstract on her own either by searching on the internet, asking Marwa during their conferences, or visiting the writing center.

These conferences offered insight into student struggles with paper in process—elucidating how the chemical engineering content drove the writing challenges as students struggled with agency, the acquisition of new knowledge, and a new way of thinking for their discipline. My postassignment interviews sought further understanding on this writing assignment as a means of entry into the discipline, with the ultimate goal of helping Dr. Zubeidi optimize it. Invitations were issued to six students who had elaborated in the open-ended questions on their surveys. Four students agreed to be interviewed. These interviews generated understanding surrounding student difficulty in meeting the criteria for the *Potential for Improvement* section as well as the limits of the assignment in terms of student investment.

Reflections

Responding to the *Potential for Improvement* Section

It was clear during both rounds of conferences that *Part Three: Potential for Improvement*, in which students were to present ideas for improving their chosen equipment, offered students the most challenge. Marwa felt the section was a bit beyond the capability of most students, largely in terms of their chemical engineering knowledge and reading level. While Marwa's assessment seemed valid, the interviews added some complexity to understanding the issue. Based on these interviews, the following possibilities emerged: that some of the students did not understand how to synthesize material to suggest improvements, that some could not figure out how to reconcile synthesizing ideas from literature with generating their own ideas, and that some were resistant to the opportunity to be innovative.

Indeed, in terms of synthesizing discipline-specific technical literature, much of the class—the 60% who had completed only FYC 2—was at a distinct disadvantage. Students who had taken FYC 3 expressed that it was necessary to have competed that course in order to respond to the research aspect of the assignment: "because in FYC 2 I still hadn't learned to research; when I got to FYC 3, that's where I learned to research and learned APA." FYC 2 required students only to synthesize 2 or 3 class readings, usually in response to a directed question; obviously, the leap between a few humanities-oriented class readings and several chemical engineering journal articles was vast. Rashid, who was taking FYC 3 simultaneously claimed he was "applying what [he] was currently learning in FYC 3 into the paper." He, too, admitted to finding the readings quite difficult, and his *Improvement* synthesis lacked substance. Surprisingly, during our interview, Rashid described an idea he had had for improving the efficiency of a bioreactor by experimenting with the size and placement of the load cells; however, he claimed he was unable to find any empirical support for his idea and decided against including it. Rashid was adamant that ideas unsanctioned by scholarship would not be acceptable, and claimed the professor led him to believe that, even as this ran contrary to the verbal and written instructions provided by Zubeidi.

Difficulties with meeting the challenge of *Part Three* were also experienced by students who had completed FYC 3. Dina, who had not only taken FYC 3 but was also taking English for Engineering, did not seem clear on the section: "I did not really understand what [Dr. Zubeidi] wanted us to do, but Marwa pointed out [during the first conference] that it was not enough to say the disadvantages of the turbines, [and that] Dr. Zubeidi would want us to improve the technique." Despite Marwa's advice during the draft stage, *Part Three* of Dina's otherwise well-written final paper did not address an improvement; it contained only the suggestion that the use of wind turbines in the UAE in far-off unpopulated regions would necessitate a means to transfer the power to the population. Mursalin, who wrote a strong paper and had synthesized well, also mentioned that, like Rashid, he had come up with original ideas that he decided not to add to the Improvements Section. Yet, Mursalin's case was different from Rashid's as he rather discounted Dr. Zubeidi's request for original ideas as fanciful. Mursalin reasoned that he simply did not have sufficient theoretical background to venture an original idea, and to do so would run counter to "thinking like an engineer."

> For example, in my case, with centrifugal pumps, when certain fluids run through the pumps, there's a possibility of corrosion. Now, I could suggest lining the pump with anti-corrosive material, but that may not be at all feasible. Maybe this anti-corrosive material doesn't exist, and even if it does, it might be costly. And there are other variables to consider.... At this point, we are at the half-life cycle of our education and we don't yet have the intellectual capacity. Yeah, you can be creative, and yeah, the world is open to you, but you can't just suggest something that seems plausible but might be utterly nonsense.

Certainly, these views shed insight into the capacities and perceptions that challenged students negotiating this section of the paper—ranging from novice writer issues and reading limitations to resistance to giving even momentary and sanctioned reign to the imagination—the imagination Dr. Zubeidi deemed necessary to engineering.

Improving the Assignment

During our first interview, Dr. Zubeidi expressed his hope that the study would indicate whether his assignment helped the students learn about their equipment and helped them appreciate the role of research in writing. Both the survey data and the conferences with the writing fellow suggested a great deal of learning was occurring, especially in relation to understanding the parts of the equipment and how they worked together: "I had never realized a compressor had a blade. I know! *A blade*!" Similarly, while acknowledging the research process and writing as challenging, students expressed appreciation for the opportunity, especially in view of the chance to revise drafts and to receive support from Marwa. These perceptions were confirmed by the four student interviews. However, two of the four—Dina and Waleed—felt that the assignment came short of its potential to engage students and to provide more of an inroad into understanding the discipline. For Dina, this response was triggered by her concurrent work in English for Engineering, a project-based course which tasks a multidisciplinary group of engineering students with an engineering-based challenge. She enjoyed the approximation of a real-life situation, which ignited her excitement for working in field, and wished Zubeidi's assignment felt similarly relevant.

Waleed, who was one of the few students who both appreciated the opportunity to suggestion innovations in *Part Three* and wrote well in that section, had taken a few humanities courses with writing assignments requiring analysis, and based on those, surmised that this assignment was too basic. Even as he admitted he learned a lot about distillation columns which he found helpful for his anticipated work in an oil refinery, he complained about the prompt: "Couldn't the professor have been more creative?" During their interviews, I invited both Dina and Waleed to brainstorm a more engaging assignment that would take into account the level of the average student in the course and would focus on a piece of equipment. In both cases, the first prompt they generated required more knowledge than could reasonably be expected. Dina's prompt—"You are a chemical engineer and you are put in a situation where a particular problem occurs with a piece of equipment; how would you handle it?"—was problematic as she realized that only students who had taken the same higher level major courses as she would be able to respond to such a prompt.

Waleed's brainstorming proved more successful once I pointed out that his suggestion to "give the student the chance to become his own boss of his own little oil refinery" was highly impractical. Like Dina, he intuitively understood the need for student investment (Norton, 2000, 2001).

> You have to make a way for the students to think of the equipment they write about as theirs. They need to be in charge of a piece of equipment. They need to care about it and to deal with it.... We learned about health and safety implications in class, so why don't we ask what could go wrong with the equipment and how would the student meet the challenge.

Agreeing that it was best to be as general as possible to give students the opportunity to interpret the assignment as per the requirements of their chosen equipment, Waleed generated the following prompt: "You are in charge of a piece of equipment for your company. Discuss what possible challenges it might present and how you could address those challenges."

Compared to the current prompt—"Discuss an equipment used in Chemical Engineering Design in detail"—Waleed's prompt offered more scope for students to engage in what Melzer (2009) refers to as an "alternative" paper which "encourage[s] exploration, synthesis, and creativity," (p. W254); moreover, it far more reflected Zubeidi's intent with the assignment.

Discussion

Earlier in this chapter appear Dr. Zubeidi's thoughts on engineering design as "open-ended [with] no correct answer [and] an infinite number of solutions." As I reflect on this study, I believe it has highlighted that Zubeidi's words apply equally to curriculum design. This is especially so in contexts that are multidimensional, where the participants—the students and professors—are, as Edward Said (1993) put it, "the political figures between domains, between forms, between homes, and between languages" (p. 332). As such, this study offers insight into the complex negotiations of a writing in the disciplines (WID) assignment at a particular Americanized liberal arts institution where social norms and pedagogical approaches are vastly different from the US and where students start university inexperienced with writing and unfamiliar with American-style academic and writing conventions. This naturalistic study locates the course research paper at the intersection of diverse societal, institutional, and educational variables that impact its existence, interpretation, and direction. This intersection varies from student to student, and indeed, can vary within each student as they begin to broaden their knowledge base and start connecting seemingly disparate identities.

This study, which chronicles both Dr. Zubeidi's efforts to integrate writing into the course and student response to those efforts, supports Herrington's (1985) conclusion that through writing assignments, professors can influence the creation of a classroom community engaged in its own distinct discipline-related discourse and practice. Herrington's conclusion has been a foundational concept of WAC and WID, as noted by Poe et al. (2010), Thaiss and Zawacki (2006), and others, and this chapter extends the scope of understanding across the globe and into a multilingual context. As we learn about Dr. Zubeidi's goals for the students—and the sociocultural and educational factors that prompt them—we understand the challenges of trying to implement WID assignments that take into account student writing and agency needs. We also see how Zubeidi, an Arab-English bilingual who was fully cognizant of how sociocultural norms and pedagogical norms had influenced his students' learning styles, was well situated to prompt his students toward goals of transfer, agency, scholarship, and creativity. Indeed, professors who understand the cultural and social implications of their context can promote learner identity and agency development (Canagarajah, 1999; Norton, 2000, 2001) as well as acknowledge and appreciate the multilingual strengths of their students. Insight from this study also adds support and complexity to the nascent scholarship initiated by Geller (2011) on the unheralded potentiality of multilingual professors in WID/WAC.

This study also extends the scholarship of writing fellow and specialist tutors into the international realm and builds upon my earlier work in the field (Ronesi, 2011). Data from my observation of the writing fellow conferences provides insight into the profound interdependence and synergy of content and writing. Marwa's ability to intervene as a "specialist"—in terms of both discipline knowledge and savviness as a seasoned chemical engineering student—enriched the quality of these conferences as she guided students toward integrating their technical knowledge into the writing, facilitated learning by asking relevant and appropriate prompting questions, and played a mentor role. The data from this study lend support to and are supported by the argument of proponents of specialist tutors or writing fellows in WID courses (Dinitz & Harrington, 2013; Gladstein, 2008; Kiedaisch & Dinitz, 1993; Soliday, 2005), particularly in engineering (Mackiewicz, 2004a, 2004b; Mackiewicz & Thompson, 2015) and is further backed by transfer literature recommending that introductory major courses with writing components be taught by disciplinary specialist (Melzer, 2014). Along that reasoning, writing fellows supporting such courses should be well versed in the disciplinary discourse conventions; this study highlights why this is especially so in a multilingual, multicultural context where the learning curve for achieving proficient writing is steeper.

The students in this study, like students at this level everywhere in the world, were vacillating between the roles enumerated earlier—those of agency avoider and agency seeker, between unknowing and newly knowing, between novice writer and proficient writer, and between student and chemical engineer. What mostly distinguishes students is the way in which context and sociocultural forces contribute to their situating of these roles, a consideration that educators must address as they design curriculum. Canagarajah aptly points out that, particularly in postcolonial discourse contexts, "[such] thinking ... has to be shaped by its own location" (1999, p. 35).

Optimizing the Writing Assignment

At the bare-bones level, the focus of this research was to optimize a writing assignment to suit the needs of the students in a second-year chemical engineering course. Subsequently, Dr. Zubeidi and I met after the semester to discuss my preliminary findings and to translate them into an improved writing assignment. Zubeidi was excited by Waleed's prompt and the reasoning behind it, and immediately decided to adopt it. To additionally support student investment, Zubeidi decided to require students to research regional companies and to understand the role their chosen piece of equipment plays within those companies. He thus added the following to the assignment sheet.

> *Introduction*—start by giving details about your position and company. Then mention why this equipment is needed for your plant operation or design. Be creative. Look up companies and decide which one you want to work for (or you are working for). Then, find out why this specific equipment is needed in your company's operation and how it will be used (i.e. you are working in pharmaceutical plant and you are in charge of a plug flow reactor).

Additionally, a number of changes were made to the assignment sheet to respond to the challenges identified by the data. To better align with the prompt, Zubeidi changed "Part Three: Potential for Improvement" to "*Possible Challenges Encountered and Their Solutions*." Concerned about the possibility of further misinterpretation about generating original and innovative solutions, he clarified by adding: "I expect your 'own' thoughts—independent of the literature—on solutions and improvements to your piece of equipment. If the paper does not contain your own thoughts in this section, the paper will not meet the requirements."

Fostering learner independence—and providing further links to the discipline—was addressed by requiring IEEE documentation only, and Zubeidi was adamant that students who were unfamiliar with the format would need to research it on their own. He would not take responsibility for teaching it, reasoning that students' foundation in APA should be adequate for figuring out IEEE. Lastly, to better align with the realities of the FYC content, Zubeidi proposed to change the literacy requirement for the course from FYC 2 as concurrent to FYC 2 as prerequisite. While this change still was not ideal—completion of FYC 3 was a sounder choice given the requirements of Zubeidi's assignment—such a requirement would have prevented students from embarking on their major course sequence at the appropriate time.

Indeed, such types of creative negotiations outlined earlier—borne of the collaboration and reflection that went into the investigation of a single writing assignment in an introductory chemical engineering course—are the foundation of reconciling American-origin curriculum with local institutional and societal cultures in order to come up with appropriate and relevant ways to support students. Naturalistic studies that elucidate the voices of the stakeholders offer the potential to tap into hidden awareness—especially in contexts where the participants negotiate various and seemingly disparate identities, and are understood as "needing help." It is not at all insignificant or incidental that students generated a new prompt and approach for the assignment that will benefit their peers. These student voices may not have had much presence in their primary and secondary education, but in view of this study, their voices need to become a significant part of their American postsecondary liberal arts education.

Appendix 1: Initial Student Survey

Dear student:

This survey is one part of a semester-long research project in which your class is involved. The focus of the research project is to learn how students understand and approach writing assignments in their university courses.

The questions in this survey will help me to understand (1) your writing experience in other university courses you have taken, (2) your awareness about writing in your field, and (3) the relationship between the writing required in your writing courses and the writing required in your discipline courses*.

* The term "discipline courses" or "discipline-based coursework" in this survey refers to **required major courses and major elective courses**—any of the courses which you have taken that advance your understanding of your field.

(1) Your name: ____________________
(2) Please check the appropriate response (only one):
__I am a Chemical Engineering major
__I am majoring in another field (please write your major in the comment box)
(3) Check the item below which reflects your current standing:
__ Sophomore 1
__ Sophomore 2
__ Junior 1
__ Junior 2
__ Senior 1
__ Senior 2
__ Other (please write your standing in the comment box)
(4) *From the following list, check the writing assignments you have been given in **discipline courses ONLY** (Again, "discipline courses" refer to **required major courses and major elective courses** only. <u>Do not include required writing in FYC courses</u>).
Check as many assignments below as applicable.
____Summary
____Journal, reflection paper
____ Informal response to a reading or class activity
____ Blogging
____ Position/issue paper
____Critique, review, or reaction paper
____Collaborative (group) written project
____Research paper
____ Lab report
____ Official letter
____ Proposal
____ Power Point/Prezi presentations
____Other (please identity the type of assignment in the comment box)
____ None of the above. I have not had writing assignments in my discipline courses.
(5) *In any of your discipline courses, have you ever read work that was written or published by the professor of that course?
___No
___Yes If yes, check (Check any that apply)

____ book
____ book chapter
____ journal article

____ conference paper
____Other (please identity the type of writing in the comment box)

(6) Based on your experience with your discipline-based coursework and knowledge about your future profession, list below **the different types of writing** that you believe you will likely engage in as a professional in your field: (i.e., memos, reports, research, etc.)
(7) *As you reflect on the different types of writing that your profession will require, how confident do you feel about your ability to write in your discipline?
(Check one.)
____Very confident
____Somewhat confident
____Not very confident
____Scared to death
(8) If, in question 7, you have answered "somewhat confident," "not very confident," or "scared to death," what practices or steps would help you increase your confidence?
(9) Which Writing course are you **currently** taking? (Check only one.)
____FYC 01
____FYC 1
____FYC 2
____FYC 3
____English for Engineering
____Other (please include in comment box)
(10) Check the Writing/English courses you have **already** taken. (Check as many as apply):
____FYC 01
____FYC 1
____FYC 2
____FYC 3
____English for Engineering
____Other (please include in comment box)
(11) What are the similarities you have noticed between Writing professors' expectations for writing in their courses and your **discipline** professors' expectations for writing in their courses:
(12) What are the **differences** you have noticed between Writing professors' expectations for writing in their courses and your **discipline** professors' expectations for writing in their courses:

*From Thaiss, C. & Zawacki, T. M. (2006). *Engaged writers and dynamic disciplines: Research on the academic writing life.* Portsmouth, NH: Heinemann.

Appendix 2: Follow-Up Survey—Student Participants

Dear student:

This survey is the second part of a semester-long research project in which your class is involved. The focus of the research project is to learn how students understand and approach writing assignments in their university courses.

The questions in this survey will help me to understand (1) your experience with the writing assignment in CHE 200 and (2) the relationship between the writing required in your writing courses and the writing required in CHE 200.

On Tuesday, January 7, you submitted a written assignment for your CHE 200 course. On the following pages are a few questions about the assignment. Kindly respond to these questions to the best of your ability.

This survey should take only 15 min of your time.

(1) What is your name? (This information will remain confidential but I need to know your name in order to contact you for a possible interview)

Below are 11 statements about the CHE 200 writing assignment you have just completed. Respond by checking one response: Strongly disagree, Disagree, Neutral, Agree, Strongly agree.

(2) This writing assignment helped me understand course content more thoroughly.
____ Strongly disagree
____ Disagree
____ Neutral
____ Agree
____Strongly agree

(3) This writing assignment helped me understand an important piece of equipment used in chemical engineering in new or more complex ways
____ Strongly disagree
____ Disagree
____ Neutral
____ Agree
____Strongly agree

(4) This writing assignment increased my general knowledge of chemical engineering.
____ Strongly disagree
____ Disagree
____ Neutral
____ Agree
____Strongly agree

(5) This assignment increased my understanding of chemical engineering as my profession (i.e., gave me more insight into my future role as a chemical engineer).
____ Strongly disagree
____ Disagree

____ Neutral
____ Agree
____Strongly agree

(6) This assignment helped to prepare me for the type of writing I will do as a chemical engineer.
____ Strongly disagree
____ Disagree
____ Neutral
____ Agree
____Strongly agree

(7) This assignment provided an opportunity to further develop my writing skills.
____ Strongly disagree
____ Disagree
____ Neutral
____ Agree
____Strongly agree

(8) My professor expected high-quality writing in this assignment.
____ Strongly disagree
____ Disagree
____ Neutral
____ Agree
____Strongly agree

(9) My choice of topic was interesting to me.
____ Strongly disagree
____ Disagree
____ Neutral
____ Agree
____Strongly agree

(10) I enjoyed working on this assignment.
____ Strongly disagree
____ Disagree
____ Neutral
____ Agree
____Strongly agree

(11) I put a lot of effort into the assignment.
____ Strongly disagree
____ Disagree
____ Neutral
____ Agree
____Strongly agree

(12) My discussions with the CHE 200 Writing Fellow helped me meet the requirements of the paper.
____ Strongly disagree
____ Disagree
____ Neutral

____ Agree
____Strongly agree

(13) List the additions/subtractions/changes you made in your paper based on your discussions with the Writing Fellow:
(14) What means of support have you used to complete this assignment? Check as many as apply.
____sought research assistance from a librarian
____used the library journal database
____used reference books in the library
____borrowed books from the library
____discussed the assignment with the professor in his office
____asked the professor questions during class
____consulted the assignment sheet the professor provided on ilearn
____examined the sample paper the professor provided on ilearn
____checked the grading rubric the professor provided on ilearn
____emailed the professor a question
____discussed my work informally with my classmates
____met and discussed my assignment with our CHE 200 Writing Fellow
____discussed my assignment with a peer tutor in the writing center
____discussed my work informally with my friends or a family member
____other (please specify)
(15) In the box below, briefly describe the aspects of this assignment that you most enjoyed.
(16) What were the challenges in this writing assignment? List them in the box below.
(17) Briefly reflect on the writing courses you have taken.
To the best of your ability, list the skills addressed in your writing classes that you applied in this CHE 200 writing assignment.
(18) What, if any advice, would you offer to your professor to improve the assignment or to better support student success in the assignment?
(19) I am extremely grateful for the time and effort you took in responding to this survey.

If you have any comments or questions regarding the CHE 200 writing assignment or my research, please share them in the box below.

References

Annous, S. A., Nicolas, M., & Townsend, M. A. (2017). Territorial borders and the teaching of writing in English: Lessons from research at the University of Balamand. In L. Arnold, A. Nebel, & L. Ronesi (Eds.), *Emerging writing research from the Middle East-North Africa region* (pp. 90–119). Boulder, CO: WAC Clearinghouse and University Press of Colorado.

Artemeva, N. (2005). A time to speak, a time to act: A rhetorical genre analysis of a novice engineer's risk-taking. *Journal of Business and Technical Communication, 19*(4), 380–421.

Artemeva, N., Logie, S., & St. Martin, J. (1999). From page to stage: How theories of genre and situated learning help introduce engineering students to discipline-specific communication. *Technical Communication Quarterly, 8*(3), 301–316.

Bilikozen, N. (2015). *Academic literacy development and identity construction of undergraduates at an American university in the UAE*. Unpublished doctoral dissertation, University of Exeter, UK.

Canagarajah, A. S. (1999). *Resisting linguistic imperialism in English teaching*. Oxford, UK: Oxford University Press.

Dannels, D. P., Anson, C. M., Bullard, L., & Peretti, S. (2003). Challenges in learning communication skills in chemical engineering. *Communication Education, 52*(1), 50–56. doi:10.1080/03634520302454.

Dinitz, S., & Harrington, S. (2013). The role of disciplinary expertise in shaping writing tutorials. *The Writing Center Journal, 33*(2), 73–98.

Driscoll, D. L. (2011, December 21). Connected, disconnected, or uncertain: Student attitudes about future writing contexts and perceptions of transfer from first year writing to the disciplines. *Across the Disciplines, 8*(2). Retrieved from http://wac.colostate.edu/atd/articles/driscoll2011/index.cfm

Geller, A. E. (2011, December 21). Teaching and learning with multilingual faculty [Special issue]. *Across the Disciplines, 8*(4). Retrieved from http://wac.colostate.edu/atd/ell/geller.cfm

Ghabra, S., & Arnold, M. (2007, June). Studying the American way: An assessment of American-style higher education in Arab countries. *The Washington Institute For Near East Policy*, Policy Focus #71.

Gladstein, J. M. (2008). Conducting research in the gray space: How writing associates negotiate between WAC and WID in an introductory biology course [Special issue on Writing Fellows]. *Across the Disciplines, 5*. Retrieved from http://wac.colostate.edu/atd/fellows/gladstein.cfm

Hall, J. (2006). Toward a unified writing curriculum: Integrating WAC/WID with freshman composition. *The WAC Journal, 17*, 5–22. Retrieved from http://wac.colostate.edu/journal/vol17/hall.pdf

Herrington, A. J. (1985). Writing in academic settings: A study of the contexts for writing in two college chemical engineering courses. *Research in the Teaching of English, 19*(4), 331–361.

Hodges, A., & Kent, B. (2017). Hybrid writing positions within WAC/WID initiatives: Connecting faculty writing expectations and MENA cultures. In L. Arnold, A. Nebel, & L. Ronesi (Eds.), *Emerging writing research from the Middle East-North Africa region* (pp. 201–215). Boulder, CO: WAC Clearinghouse.

Hughes, B., & Hall, E. B. (Eds.). (2008, March 29). [Special issue on Writing Fellows]. *Across the Disciplines, 5*. Retrieved from http://wac.colostate.edu/atd/fellows/index.cfm

Johnson, J. P., & Krase, E. (2012, October). Coming to learn: From first-year composition to writing in the disciplines. *Across the Disciplines, 9*(2). Retrieved from http://wac.colostate.edu/atd/articles/johnson_krase2012.cfm

Kiedaisch, J., & Dinitz, S. (1993). Look back and say 'so what': The limitations of the generalist tutor. *The Writing Center Journal, 14*(1), 63–74.

Knowledge & Human Development Authority. (2010/2011). *Arab knowledge report 2010/2011: Preparing future generations for the knowledge society*. Retrieved from http://www.arab-hdr.org/akr/index.aspx

Leydens, J. A. (2012). Sociotechnical communication in engineering: An exploration and unveiling of common myths. *Engineering Studies, 4*(1), 1–9. doi:10.1080/19378629.2012.662851.

Mackiewicz, J. (2004a). Hinting at what they mean: Indirect suggestions in writing tutors' interactions with engineering students. *IEEE Transactions on Professional Communications, 48*(4), 265–276.

Mackiewicz, J. (2004b). The effects of tutor expertise in engineering writing: A linguistic analysis of writing tutors' comments. *IEEE Transactions on Professional Communications, 47*(4), 316–328.

Mackiewicz, J., & Thompson, I. K. (2015). *Talk about writing: The tutoring strategies of experienced writing center tutors*. New York: Routledge.

McNair, L., & Venters, C. (2012, June). *Work in progress: Using writing-to-learn methods to improve conceptual knowledge in Engineering Statics*. Paper presented at IEEE Frontiers in

Education Conference, Seattle, WA. Retrieved from http://ieeexplore.ieee.org/xpls/abs_all.jsp?arnumber=6462525&tag=1

Melzer, D. (2009). Writing assignments across the curriculum: A national study of college writing. *College Composition and Communication, 61*(2), 240–261.

Melzer, D. (2014). The connected curriculum: Designing a vertical transfer writing curriculum. *The WAC Journal, 25*, 78–91. Retrieved from http://wac.colostate.edu/journal/vol25/melzer.pdf

Miller, R. T., & Pessoa, S. (2017). Integrating writing assignments at an American branch campus in Qatar: Challenges, adaptations, and recommendations. In L. Arnold, A. Nebel, & L. Ronesi (Eds.), *Emerging writing research from the Middle East-North Africa region* (pp. 176–200). Boulder, CO: WAC Clearinghouse.

Miller-Idriss, C., & Hanauer, E. (2011). Transnational higher education: Offshore campuses in the Middle East. *Comparative Education, 47*, 181–207.

Mullin, J., Schorn, S., Turner, T., Hertz, R., Davidson, D., & Baca, A. (2008, March 29). Challenging our practices, supporting our theories: Writing mentors as change agents across discourse communities [Special issue on Writing Fellows]. *Across the Disciplines, 5*. Retrieved from http://wac.colostate.edu/atd/fellows/mullin.cfm

Nebel, A. (2017). Linguistic superdiversity and English-medium higher education in Qatar. In L. Arnold, A. Nebel, & L. Ronesi (Eds.), *Emerging writing research from the Middle East-North Africa region* (pp. 33–45). Boulder, CO: WAC Clearinghouse.

Norton, B. (2000). *Identity and language learning: Gender, ethnicity, and educational change*. New York: Longman.

Norton, B. (2001). Non-participation, imagined communities and the language classroom. In M. Breen (Ed.), *Learner contributions to language learning: New directions in research* (pp. 159–171). Harlow, England: Pearson Education.

Patton, M. D., & Taylor, S. S. (2013, May 21). Re-evaluating directive commentary in an engineering activity systems. *Across the Disciplines, 10*(1). Retrieved from http://wac.colostate.edu/atd/articles/patton_taylor2013.cfm

Peters, D. L., & Arbor, J. (2014, June). *Student perceptions of connections between statics class and co-op work experience*. Paper presented at ASEE Annual Conference, Indianapolis, IN. Retrieved from http://www.asee.org/public/conferences/32/papers/8496/view

Pilotte, M. K., Bairaktarova, D., & Evangelou, D. (2013). Trans-discipline engineering communication and norms: An exploration of communication behaviors within engineering practice. *Australasian Journal of Engineering Education, 19*(2), 87–99.

Poe, M., Learner, N., & Craig, J. (2010). *Learning to communicate in science and engineering: Case studies from MIT*. Cambridge, MA: MIT.

Prescott, D., El-Sakran, T., Albasha, L., Aloul, F., & Al-Assaf, Y. (2011). Engineering communication interface: An engineering multi-disciplinary project. *US-China Education Review, A*(7), 936–945.

Ronesi, L. (2011, December 21). "Striking while the iron is hot." A writing fellows program supporting lower-division courses at an American university in the UAE. *Across the Disciplines, 8*(4). Retrieved from http://wac.colostate.edu/atd/ell/ronesi.cfm

Rudd, M., & Telafici, M. (2017). An Arabian gulf: First-year composition textbooks at an international branch campus in Qatar. In L. Arnold, A. Nebel, & L. Ronesi (Eds.), *Emerging writing research from the Middle East-North Africa region* (pp. 120–136). Boulder, CO: WAC Clearinghouse.

Sageev, P., & Romanowski, C. J. (2001). A message from recent engineering graduates in the workplace: Results of a survey on technical communication skills. *Journal of Engineering Education, 90*(4), 685–692.

Said, E. (1993). *Culture and imperialism*. New York: Alfred Knopf.

Severino, C., & Traschel, M. (2008, March 29). Theories of specialized discourses and writing fellows programs [Special issue on Writing Fellows]. *Across the Disciplines, 5*. Retrieved from http://wac.colostate.edu/atd/fellows/severino.cfm

Shine, E. A. (2008). *Written feedback in a freshman writing course: Instructors' and students' perspectives on giving, getting, and using feedback*. Unpublished doctoral dissertation, Massey University, New Zealand. Retrieved from http://hdl.handle.net/10179/3947

Shine, E. A. (2013). Reading experiences and avoidance in freshman classes. In *European Conference on Education 2013 Proceedings*. Retrieved from http://iafor.org/archives/offprints/ece2013-offprints/ECE2013_0332.pdf

Soliday, M. (2005). General readers and classroom tutors across the curriculum. In C. Spigelman & L. Grobman (Eds.), *On location: Theory and practice in classroom-based writing tutoring*. Logan, UT: USU Press. Retrieved from http://digitalcommons.usu.edu/usupress_pubs/151

Soven, M. (2001). Curriculum-based peer tutors and WAC. In S. H. McLeod, E. Miraglia, M. Soven, & C. Thaiss (Eds.), *WAC for the new millennium. Strategies for continuing writing-across the curriculum programs* (pp. 200–223). Urbana, IL: NCTE.

Spigelman, C., & Grobman, L. (Eds.). (2005). *On location: Theory and practice in classroom-based writing tutoring*. Logan, UT: USU Press. Retrieved from http://digitalcommons.usu.edu/usupress_pubs/151

Thaiss, C., & Zawacki, T. M. (2006). *Engaged writers and dynamic disciplines: Research on the academic writing life*. Portsmouth, NH: Heinemann.

United Nations Development Programme. (2003). *UNDP Arab human development report 2003: Building a knowledge society*. New York: United Nations Publications. Retrieved February 18, 2010, from http://www.arab-hdr.org/publications/other/ahdr/ahdr2003e.pdf

Winsor, D. A. (1996). *Writing like an engineer: A rhetorical education*. Mahwah, NJ: Lawrence Erlbaum Associates.

Creativity and Identity in the Construction of Professional Portfolios

Lisa D. McNair, Marie C. Paretti, and Christopher Gewirtz

Introduction and Background

The concept of identity construction as a creative act is rooted in a broader social constructionist conception of identity (Abes, Jones, & McEwen, 2007). This approach, as Bamberg, De Fina, and Schiffrin (2011) note, differs from more traditional views that posit identity as an essence that can be discovered. In particular, we follow Bamberg et al. in treating identity as something that is "discursively constructed," or made through language. This perspective offers researchers a means

> to study identity as constructed in discourse, as negotiated among speaking subjects in social contexts, and as emerging in the form of subjectivity and a sense of self. Our suggestion [to study identity in this way] implies a shift away from viewing the person as self-contained, *having* identity, and generating his/her individuality and character as a personal identity project toward focusing instead on the processes in which identity is *done* or *made*—as *constructed* in discursive activities. (Bamberg et al., 2011, p. 178)

At the same time, by viewing identities as constructed through language, this approach offers space for educators to intentionally and effectively support students in this creative process. Professional portfolios offer one approach to providing such support. Rather than treating engineers as a monolithic group characterized solely by skill in math and science, portfolios offer educators a way to help students creatively reimagine themselves and their field. The digital portfolios in this study, we argue, illustrate the ways participants created public representations of themselves as engineering professionals, and at the same time created their own definitions of what it means to be an engineer.

The categories of self as researcher, teacher, and community member lend themselves to frameworks that treat identity not as a single essence but rather as a complex of identities that individuals flexibly integrate and deploy based on context

L.D. McNair (✉) • M.C. Paretti • C. Gewirtz
Department of Engineering Education, Virginia Tech, Blacksburg, VA, USA
e-mail: lmcnair@vt.edu

D. Bairaktarova, M. Eodice (eds.), *Creative Ways of Knowing in Engineering*,
DOI 10.1007/978-3-319-49352-7_6

and meaning making. Abes et al.'s (2007) model of multiple dimensions of identity captures this perspective. Their work explores the ways in which various components of identity (e.g., not only demographic categories such as race, ethnicity, and sexual orientation, but also social categories such as class, religion, or profession) intersect with one another and, equally important, shift in salience as individuals move from one context to another. In a very simplified example, gender and profession may be the most salient components of a person's identity in one social context, while race and religion may be most salient in another.

Moreover, Abes et al. argue that these complex intersections of identity dimensions are shaped by individuals' capacity for meaning making, or the ways in which they make sense of the world. To address the role of meaning making, they introduce Baxter Magolda's concept of self-authorship (Baxter Magolda, 1992; Baxter Magolda & King, 2004) into their model. Baxter Magolda's model posits college students' development in terms of a move from formulaic, right/wrong, authority-based models of knowledge to the recognition of multiple possible truths, and ultimately to a critical stance in which the knower weighs, evaluates, and negotiates these multiple truths intentionally and with agency. It is this final stage that Baxter Magolda terms "self-authorship." In terms of multiple dimensions of identity, Abes et al. argue that self-authorship serves as a filter that enables individuals to consciously recognize, define, integrate, and deploy multiple dimensions of their identity. That is, individuals on the formulaic end of Baxter Magolda's spectrum accept external definitions of facets such as race, profession, or religion, and accept external valuations of whether and how those facets are compatible. Individuals who are self-authors, in contrast, reshape those external definitions to suit their own needs and interests, integrate facets that others might deem incompatible (e.g., woman and engineer) and can actively and intentionally perform or foreground different facets based on context. Self-authorship thus represents a creative act of identity construction.

Much of the work on identity construction in education broadly and in engineering specifically has focused on undergraduate students (e.g., Abes & Jones, 2004; Abes et al., 2007; Capobianco, 2006; Foor, Walden, & Trytten, 2007; Jones, 2000; Pierrakos, Beam, Constantz, Johri, & Anderson, 2009; Tate & Linn, 2005; Tonso, 2007, 2014), though some work exists on graduate students (e.g., Crede & Borrego, 2012; Crede, Borrego, & McNair, 2010). Often, however, these studies look at the ways students' identities develop "in the wild," so to speak, without interventions to support students in this process. Yet as promising work by Turns & Lappenbusch (2006) and others suggests, interventions such as portfolio construction can serve as important scaffolds to provide students a platform from which to engage in intentional, creative self-construction. In this context, graduate engineering students are a particularly interesting population to consider. As Baxter Magolda notes, few undergraduates achieve self-authorship by the end of four years in college; graduate students, however, are further along their professional trajectories and are immersed in the uncertainties and ambiguities of research, which may increase their capacity for self-authorship. They also typically bring life experiences, research interests, professional interactions, and career plans that are likely to coincide with increased self-knowledge. Thus, we suggest that graduate students may have greater meaning making capacity, and may accept both the multifaceted nature of knowledge and the need for a critical stance to negotiate among those facets intentionally.

Portfolio to Professoriate: The P2P Project

To explore graduate students' creative identity constructions, we draw on data from an NSF-funded project designed to encourage reflective practice in graduate students who are constructing professional identities as both researchers and educators (McNair & Garrison, 2013; Svyantek, Kajfez, & McNair, 2015). We have developed a digital portfolio process, Portfolio to Professoriate (P2P), that guides students through building a portfolio and encourages their development of integrative thinking. P2P is a curriculum for graduate students in STEM, with an emphasis on engineering students. The P2P curriculum divides the process of creating a professional digital portfolio into weekly tasks. As the project has evolved, the curriculum has been redesigned to be more flexible and self-directed. At each campus, students have been offered the opportunity to engage in peer feedback—sometimes online and other times in person. As of this publication, the P2P curriculum has been piloted on five campuses under different circumstances, using different incentives, as shown in Table 1:

To create their professional portfolios, students were asked to upload evidence of and write narratives about their accomplishments in four categories: (1) Research, (2) Teaching, (3) Service, and (4) Lifelong Learning. The first three reflect the primary components of faculty work, while the fourth represents a tool to help participants engage reflectively in their own process of career development.

Typically, participants uploaded published papers or slide presentations used during conference presentations as evidence under Research, along with a variety of visuals (e.g., charts, illustrations, photos, and even a few videos) depicting their research. Under Teaching, the typical evidence uploaded might be anything from a syllabus or a test they created to a teaching philosophy or a video of them actually teaching; several also uploaded course evaluations and comments from their students. Under Service, participants often had multiple pieces of inspiring evidence, such as starting Engineers Without Borders chapters, reviewing of conference papers for a professional organization, or mentoring undergraduate engineering students. But the most common service activities involved working with K12 students in an engineering context. Such activities ranged from judging statewide robotics competitions to teaching unprivileged 5 and 6 year olds how to build a bridge during summer camp. Under Lifelong Learning, learning a language and advanced mastery of a musical instrument were popular choices, as was pursuing a graduate certificate in teaching or another STEM field.

Each piece of evidence students upload typically documents a specific professional accomplishment. Students are then asked to write narratives to pair with their uploaded evidence. These narratives not only provide helpful background on the circumstances of the accomplishment, but also speak to what the student learned through the process.

Table 1 Summary of participating sites

Institutional circumstances							Participants	
Site	Term	Research-intensive institution	Stipend	Course credit	Required by Dept. for Assessment	Other	Number of students signed up to participate in P2P	Number of students completing a portfolio (completion rate)
A	AY 2011–2012	Yes	Yes	No	No	All volunteers, two in-person meetings	11	8 (72 %)
B	Winter—Spring 2012	Yes	No	No	No	All volunteers, participants part of a cohort related to another topic, met weekly	16	2 (12 %)
C	Spring 2012	Yes	No	No	No	All volunteers, all participants part of course on teaching	9	1 (11 %)
D	Spring 2012	Yes	No	No	Yes	All volunteers	4	2 (50 %)
E	Fall 2012	Yes	No	Yes	No	All students awarded 20 % of a course grade for completing portfolio according to a rubric standard	6	6 (100 %)
Total							46	19 (41 %)

In addition, during the P2P curriculum, students were asked to create a Home page that welcomes readers and explains the contents of the portfolio, along with an About page that typically displays a short bio, a photo, contact information, and—often—a link to their CV. Several portfolios also included philosophies of engineering and/or reflections on participants' career development.

Methods—Qualitative Coding for Integrative Thinking

Participants

Graduate students in engineering disciplines from four research universities agreed to participate in the P2P curriculum as part of our research study. The study included taking pre/post-surveys on role identity, constructing a digital portfolio, and participating in an interview or focus group. The pool of participants made up five cohorts differentiated by individual contexts and the following factors: stipend, incentive, required for course credit, and required for assessment (Table 1). A total of 46 students signed up for the study, and 41 % of students who began the study completed full portfolios (defined as at least one piece of evidence and corresponding reflection in each of the four categories). All participants signed IRB-approved consent forms.

Data Collection

This chapter focuses on the digital portfolios themselves. Because participants own and control their digital portfolios (and thus could further develop or destroy them after completing the P2P program), the portfolios were archived offline at the end of the curricular period for each cohort to ensure stable, consistent objects for analysis. As indicated above, the portfolios included images, documentation of work (e.g., presentation files, documents, images, and videos), textual narratives, and links among internal site pages as well as links from the portfolio pages to external sites (note that we preserved the links themselves, but did not attempt to archive external sites). All of these elements were viewed as parts that work together to create units of meaning. For example, a piece of documentation is "annotated" by a student's written narrative that explains the context of the documentation and makes connections to the student's professional roles.

In addition, researchers wrote case descriptions of each student, in which they noted patterns or common threads throughout a student's portfolio reflections; adherence to, consistency in, and ownership of portfolio process; response to feedback from both faculty and peers; and quality of portfolio according to criteria such as writing, visuals, navigation, and audience.

Table 2 Identity integration codebook (McNair & Garrison, 2013)

Roles	
Teacher	Imparting knowledge or giving instruction
Researcher	Creating new knowledge. Investigation
Volunteer	Unpaid work for the benefit of others
Student	Formal role, being in the classroom. Structured activities. Obligatory
Learner	Pursues learning in many contexts. Self-directed and self-motivated
Professional	Use this code for roles like engineer, designer, etc.
Qualities	
Lifelong learning	Drive for learning that takes place inside and outside the classroom. Past knowledge used in new situations
Curiosity	Drive for knowledge and exploration. Inquisitive
Altruism	Interest in or drive toward improving the lives of others
Creativity	Artful. Using imagination. Developing an original idea or solution
Degree of integration	
Overlap	Loose connection. Loose interaction. Accidental
Integrate	Deep connection. Experiences from one role influencing another role. Fused together. Intentional

Data Analysis

A coding scheme for the full data set was developed iteratively as detailed in McNair and Garrison (2013). In brief, three researchers independently open-coded (Patton, 2002) one student portfolio around the required portfolio categories (teaching, research, service, and lifelong learning), instances of overlap and integration across these categories, and emergent codes. Together with a fourth researcher, the team compared codes and established a codebook (shown in Table 2) that included roles, qualities, and degree of integration.

Three coders then reanalyzed the initial portfolio, iteratively comparing coding choices to reach consensus on the application of the codes; these coders then coded the remaining portfolios independently. In prior work, McNair and Garrison (2013), compared the coding to the case descriptions to identify patterns regarding (a) the presence of integrative thinking, (b) links between specific portfolio categories (teaching, research, service, lifelong learning) and integrative thinking, and (c) factors that encouraged engagement in the portfolio process. With respect to links between portfolio categories and integration, the coders also used the category "Miscellaneous" to capture text from sections of the portfolio such as a Home page, an About page, narratives written on career interests during childhood, and philosophy statements about their field or professional broadly.

That analysis showed that across the data set, integration was evident in each portfolio category and appeared in each portfolio, though not every participant showed integration in every category. Notably, integration occurred most often in Service, Lifelong Learning, and Miscellaneous. In part, this finding may reflect the fact that Teaching and Research are both well defined and potentially more familiar to most graduate students, whereas Service, Lifelong Learning and Miscellaneous

Table 3 Self-authorship codebook

Meaning making (degree of self-authorship)	
External formulas	The responsibility for meaning making is given to others, often authorities. Relies on external definition of one's identities. Reflections contain rationalizations or cognitive dissonance about one's intersecting identities, and do not address conflicts between those identities
Crossroads	The intermediary step between external formulas and self-authorship. Often, tensions between external formulas and internal values are described, as one begins to discover limitations of external formulas
Self-authorship	High capacity for meaning making, which allows for negotiation and integration between identities. There is no reliance on stereotypes, but rather the core values or "internal voice" of the individual are trusted

are more open-ended, affording more opportunities for creativity and reflection, as well as less familiar, potentially encouraging more creativity (McNair & Garrison, 2013). In addition, "Miscellaneous" included sections in which students described things like their philosophies of engineering or their personal journeys (requirements varied by portfolio site), components that also invite integration.

To more fully explore these findings and examine in detail how the participants created professional identities, a fifth researcher reviewed the full data set and identified nine portfolios that were complete (i.e., evidence and reflection in each category, along with Home and About pages) and showed evidence of integration (as defined in Table 2) across at least two categories. All portfolios meeting these criteria were included in the present analysis. The original researcher and two new researchers reviewed this subset, including the prior coding, focus group findings, and case descriptions. Consistent with Abes et al.'s (2007) multiple dimensions of identity, the research team also reviewed the portfolios to identify the dimensions of identity present in each portfolio and the degree of self-authorship. The dimensions of identity included not only the roles listed in Table 3, but also dimensions specific to each individual (e.g., ethnic heritage, family status). With respect to self-authorship, the researchers applied the definitions in Table 3.

The results were used to construct identity narratives for each of the nine participants. These narratives are reported in the following section through an interpretivist framework that emerged as the research team iteratively reviewed dimensions of identity, sites of integration, and elements of self-authorship within and across the nine cases. We then situate emergent patterns within theoretical frameworks of identity construction as a creative act (Abes & Jones, 2004; Bamberg et al., 2011; Baxter Magolda, 2004).

Findings: Narratives of Creative Identity Construction

Following Judith Butler, Abes et al. (2007) recognized identity as performative and fluid; these portfolios are such performances, highlighting multiple facets of each participant's identity. We describe each portfolio first in terms of the dimensions of identity participants offer and their ways of talking about themselves and about engineering. We then describe the primary thread or threads that serve as foci of integration across portfolio components, and highlight instances of creative self-authorship. We note race or ethnicity only when a participant self-identified in the portfolio. With respect to gender, in five cases participants used a third person pronoun (e.g., on their "About" page or in reporting others' comments about them) that identified a gender, and we followed this lead. In two cases, participants reported involvement with an organization that suggested gender (Lance is a Boy Scout troop leader and Lindy participates in a Women in Engineering group). Corbin and Denver make no reference to gender; in these two cases, we used masculine pronouns based on name and physical appearance.

Corbin

Corbin is a young man of Middle Eastern heritage who includes a small head shot of himself in an open collared shirt on his Home page, but offers few details of his personal life or educational experiences. His understanding of engineering moved from stereotypical childhood strengths in math and science to a richer understanding of engineering as the application of these and other domains "to solve problems and aid humanity." He then positions himself as "a contributor to the global community of engineers" working toward this goal, with his research focused specifically on sustainable energy. His discussions of service and lifelong learning pick up on a similar theme (aiding humanity), but do so in terms of his cultural heritage—the one personal dimension he does include. His discussions of service and his lifelong learning focus on bringing together Israeli and Palestinian students in dialogue to end the "partisan atmosphere," and he includes his own poetry on his Middle Eastern family roots.

One key thread in Corbin's portfolio, then, is the use of communication as a means to "aid humanity." With respect to research, this thread emerges as he talks about positioning his research writing to enable him to contribute to that "global community of engineers." In teaching, he emphasizes his capacity to listen as he describes how his work as a grader helped him develop "unique knowledge of each student's performance on each assignment" that he then used to structure review sessions. His use of communication to process, understand, and act is perhaps most visible in his efforts to promote peaceful Israeli–Palestinian dialogue on his campus. He emphasizes the need to bring in multiple perspectives, and the events he organized also bring in multiple modalities, including a teach-in, a film, a talk with representatives from opposing sides, policy writing, and other communicative acts that promote understanding and exchange of ideas.

Corbin's engagement with the Israeli–Palestinian discourse also illustrates his capacity for self-authorship. In his writing, he explicitly rejects externally defined portrayals of his family's native country in "evening news stories." Instead, he seeks to understand and engage with people from a country at war with his own, working to provide a safe space for students from all involved nations to communicate with each other. As he explains, "I learned to silence some of my own dogmatic principles and bias in order to understand how someone different from me feels about an issue that is mutually significant to us." He thus moves from nationalistic partisanship to constructive engagement, intentionally reinscribing his own position and creating a new space for dialogue.

Denver

Denver is an African-American male whose Home page includes photos of himself backpacking and scuba diving though no text in his portfolio further explains those activities. He describes his research as operating at the intersection of "[biomechanical devices] and the human user" and sees engineering as a means of "advancing the human condition." He positions himself as a role model, recognizing that "other students are looking at me" implicitly because of race. In the same section of the portfolio, he describes wanting to build research collaborations with colleagues at Historically Black Colleges and Universities (HBCUs) and other minority serving institutions, and in a later section he highlights his commitment to expanding the racial diversity of the field and supporting students and faculty of color. His service activities also focus on minority students in STEM fields as he details his efforts to help provide a platform to bring together minority graduate students, faculty, and staff—the first such effort on his campus. In addition to highlighting his racial identity, he also positions himself as interdisciplinary, highlighting his background in physics, his interest in photography, and, in his research, his exploration of qualitative methods to understand user experiences.

One pervasive thread in Denver's portfolio is his emphasis on altruism. This altruism is evident not only in his understanding of engineering as a means of "advancing the human condition," but also in his teaching as he searches for principles that apply to all learners, in his research as he embraces qualitative methods to learn more about how technologies interface with users, and especially in his extended service discussion about bringing individuals of color together across academic hierarchies.

With respect to self-authorship, Denver's portfolio not only presents but also emphasizes his position as a person of color in engineering. Both as a visible role model and an active supporter of minority students and faculty, he intentionally chooses to bring together his identities as African-American and engineer, and to use his visible presence to help recreate the demographic profile of the field. Moreover, his altruistic desire to connect people and technology pushed him to develop skills that are "completely outside of [his] discipline," such as qualitative observation. In addition, he exerts agency over his research and teaching activities as he describes learning both how to bring his passions into these spaces and how to present them in ways that are productive. As a researcher, for example, he notes that he is "not interested in just making incremental changes; rather [he wants to make] groundbreaking revolutionary changes that affect the current thoughts and opinions of the field at large."

Edwin

Edwin introduces himself as a "chemical engineer," "researcher," and "educator," and his Home page shows a headshot of him in a casual short-sleeve shirt. He also includes a picture of himself with a female partner, a baby, two dogs, and sled in the snow on his About page, though he offers no comment about this photo other than to note the sled is "home-made." He positions engineering "at the intersection of fundamental physics and the products we all use." As a chemical engineer, in particular, he emphasizes the human and environmental hazards posed by toxic chemicals and focuses much of his work on reducing these hazards. His discussion of services focuses on his professional contributions to lab safety in his university. His portfolio is characterized by an extensive collection of personal stories—of his childhood, of building projects (including work with Habitat for Humanity), and hands-on engagement in his teaching, research, and engineering work. He uses humor throughout his reflections to personalize his discussions and create an engaging sense of self. At the same time, he includes virtually no interests, hobbies, or experiences outside engineering—his service, learning, teaching, and research reflections all focus on engineering. His portfolio is thus dominated by a single dimension of his identity though one that unfolds over time through a variety of reflective experiences.

A dominant thread in Edwin's portfolio is this process of learning through reflective practice. In his introductory reflection, he describes an experience of a failed project in which he lost a treehouse to a windstorm. As he "stood staring at the three trunks," noticing that the trees swayed in different directions, he decided in his next iteration to use two floating point supports to allow for motion. Using this early instance of reflection as an extended metaphor, he cites reflection as a tool for perseverance: "With each new challenge I encounter I can always think back to the tree-house that fell down and know that a solution is out there, I just need to stare at the trees." This practice, in which he learns from experimentation without fear of failure, recurs throughout his portfolio. In his research, for example, he describes a side project he is pursuing because he expects "interesting" results and is eager to continue "digging"; he describes his thesis as the result of "pulling pieces of information from seemingly unrelated fields and putting them together to see what you get." In his teaching, he continuously asks students for feedback so that he can iterate on his teaching methods. In the section on lifelong learning, he again cites failure as an opportunity and points to awareness of context as a way to both deepen and broaden his understanding of the world. Through these experiences, both his tone and his descriptions display curiosity, playfulness, openness to multiple solutions, and an appreciation for failures as learning opportunities.

Because Edwin highlights only one dimension of his identity, it is more difficult to characterize his level of self-authorship across multiple domains. His portfolio does, however, explicitly integrate his work in research, teaching, service, and learning, and he constructs a consistent image of himself as a curious, playful learner. In describing his own development over time, moreover, these learning experiences emphasize his ability to listen to that "internal voice" central to self-authorship and to narrate his story into an integrated professional self.

Fraser

Fraser's Home page includes a head shot of himself in a button-down shirt and tie, while his About page describes his love of sports, including both running and sailing, and shows a photo of him waving a foam hand in a sports stadium rooting for his favorite team. He sees engineering as the process of "solving societal problems by artfully applying math and science through design," and he emphasizes the ways in which both the artfulness and the social dimensions emerged for him over time as he progressed through his education. He includes discussions of learning Spanish as well as pursuing interdisciplinary interests in business and marketing; in fact, he characterizes himself as interdisciplinary, noting that he "had a hard time deciding on a major in college" and that he finds "life is more interesting when there's variety." Like Edwin, his writing adopts a personalized, storytelling tone, and like Corbin, he includes activities associated with his cultural/religious heritage: his CV includes several activities to support and build community for young Jewish professionals. Unlike Corbin, however, he does not discuss or describe these activities in his reflections.

Like Edwin, Fraser's portfolio includes a strong learning thread, but in contrast to Edwin's reflective learning from failures, Fraser emphasizes strategic inquiry, noting his love of strategy games, including both physical sports and various board games. His sense of strategic exploration is reflected in his childhood "technical curiosity" about how things work and in his description of himself as an adult lifelong learner who "hasn't stopped asking questions" and "will never stop learning." This strategic inquiry characterizes his teaching practice as he models curiosity by posing questions and encouraging students to come up with ideas that interest them. Similarly, his dissertation research focuses on a new combination of ideas from different areas that he brings together by drawing on systems thinking. Even his approach to service is strategic, as he explains, "Service is an integral part of my life, not because a guilty conscience tells me to do it, but because I've found ways to make service meaningful to myself."

Fraser's emphasis on making meaning for himself reflects one of the ways self-authorship emerges in his portfolio. He notes, in fact, that he has never been able to settle on "favorites" because so many things in the world are interesting at different times; that is, he resists the idea that he should fix his choices and preferences, and instead exerts agency in noting that his favorites change over time and with context. Importantly, he does not offer this observation as a sign of instability or whimsicality;

rather, it reflects the same kind of intentional strategic thinking that characterizes his story of personal development. His ability to connect various threads in his life (including, for example, using statics to explain sailing to friends, or linking the strategy of sports to the systems thinking he brings to engineering) also points to evidence of self-authorship.

Gloria

Gloria's About page includes a neutral headshot of herself, but her Home page has a centered photo of a silhouetted kayaker on the ocean at sunset, and she uses the metaphor of a journey to describe her personal and professional development. Her philosophy of engineering begins in problem solving, but layers on critical thinking and connection to the world. Drawing on her own industry experience, she emphasizes the ways in which engineering is "application driven" and thus practical, but also the ways in which the work engineers do requires them to "think 'outside the box'" in ways not often taught in school. She thus emphasizes the importance of mentoring both in and beyond school in developing new engineers. Her service includes mentoring others in a number of K-12 engineering outreach programs as well as forming a chapter of Engineers Without Borders (EWB) at her undergraduate institution. But she also describes her work with several church and community organizations, and in illustrating lifelong learning, she focuses on her passion for learning French both in school and in a semester abroad. Notably, Gloria's is one of the most visually rich portfolios in the data set in that it contains over two dozen photos, including photos of herself in a variety of engineering contexts (labs, EWB projects) as well as photos of class activities and groups she has worked with and photos of her travels that echo her metaphor of a journey.

The breadth and diversity of Gloria's portfolio make it difficult to identify a single strong thread, and in fact, her metaphor of a journey and her inclusions of images such as the kayaker on her Home page or the path through the woods next to the narrative of her professional journey highlight both directionality and change. Two facets emerge as particularly salient in this journey however: practicality and collaboration. Her sense of practicality emerges in the ways in which she talks about her industry experiences and her perception of engineering as application oriented, but also in her sense of herself as a "planner" and "analyzer"—qualities evident across her descriptions of various activities. At the same time, through both the numerous photographs of groups of people and her explicit reflection on "the importance of collaboration," she emphasizes collaborative work in each of the dimensions of her life—learning from mentors in industry, working with other researchers, and organizing groups in her local communities.

As suggested by her explicit use of the journey metaphor and the inclusion of a wide array of different dimensions of her identity, Gloria's portfolio demonstrates several different elements of self-authorship. She characterizes herself as a learner (and even includes a section in her portfolio titled "Learning for the Sake of

Learning") and highlights the ways in which various learning experiences help her understand more about what she does and doesn't want to do professionally and personally. For example, both her undergraduate work and her industry experience show her that she "cannot envision [herself] continuing down the path of a design engineer," and instead take her into an academic career. Moreover, as both a learner and a student, she recognizes the value of not only learning from teachers but "the eventual fading of dependence on others and an elevation in confidence for the given task" and she wants to be "proficient" in teaching techniques that help students progress as "self-learners," all of which point to the type of critical synthesis indicative of self-authorship.

Isabella

Isabella's Home page includes a "fun" picture of herself holding a box of Nerds candy, and in her About Me section, she describes growing up in Asia until age 5 and wanting to move somewhere warm. Her philosophy of engineering focuses on the idea of "dream advancement," and she sees her field of computer science as one that helps integrate technologies into our lives in ways that enable the advancement "of not only my dreams, but of others." Her section on lifelong learning includes an extensive discussion of contract bridge, a hobby she is passionate about and one she readily links to but also separates from her field: "Academically, this is a resource optimization and probability problem. Socially, it's just fun ☺." Her reflections on service focus on her efforts to help other teachers in her university by working as a teaching consultant.

A core thread in Isabella's portfolio is teaching and learning: she sees sharing her knowledge with others as paramount though she is uncertain about her long-term vision for that work (e.g., research-intensive versus teaching-intensive institution). Isabella states that she has always had difficulty learning, and even describes herself as slow to "master or even pick up a new subject matter"; yet she has earned degrees from highly ranked institutions and has received international awards. Her reflections narrate learning trajectories that focus on gains in confidence, flexibility, and contextual awareness; at the same time, these gains allow her to become comfortable with uncertainty as she considers her future directions. Her reflections track her transition from someone with little confidence to someone who recognizes her own accomplishments and is confident in her ability to guide a project from start to finish. For example, she notes that completing a research project taught her about "what research is, what it means to be a researcher, and the standards of creating 'good' research." She constructs her teacher evolution as moving from not knowing what to do and feeling extremely challenged to confidently mentoring others.

Interestingly, with respect to self-authorship, Isabella also describes herself as someone who is motivated by "external validation." But as she narrates her development as a teacher in particular, she explains how her understanding of that validation shifted from believing that students "have to like you, and if they liked you, they would listen to you," to "figure[ing] out how to say 'no' and not be afraid they'll hate me." She ultimately finds validation in her students' learning rather than their approval. In doing so, she is moving increasingly toward self-authorship in which she understands herself based not on an external formula that she has lived up to, but rather based on her ability to, as Baxter-Magolda would call it, "trust her internal voice" and recognize that effectively engaging students in learning is more valuable than simple approval. At the same time, Isabella still acknowledges the importance of students' feedback and includes almost two dozen positive comments from her peer and student evaluations in her portfolio; her portfolio thus positions her at the crossroads in her journey toward self-authorship.

Jazmin

Jazmin places a casual photo of herself on a speedboat on her Home page and a photo of herself with her partner and a dog on her About page. Her portfolio is marked by a strong emphasis on collaboration, particularly across disciplinary boundaries. Her vision of engineering begins with the stereotypical "engineers as problem-solvers" but expands to include not only "problem-solving within constraints," but also imagination, invention, and creativity. Importantly, like most of the participants, she also explicitly positions engineering in terms of an altruistic goal: "to help others and create a better society." Moreover, her conception of engineering is holistic in that she sees engineers as being "involved in nearly every aspect of society." Her sections on lifelong learning and her professional journey present several different dimensions of her identity, including a range of liberal arts courses she has taken that have helped shape her personal and professional goals as well as her learning in a variety of personal pursuits such as knitting, cooking, and fitness.

A central thread in Jazmin's portfolio is working across disciplines and with diverse collaborators. She notes, for example, that her dissertation involved "a complex, interdisciplinary project [that] really highlighted the value of teamwork," and she reflects on the ways her postdoctoral position has helped her learn to "giv[e] up control and [let] others be successful with experiments." She also sees service as a way to "build networks and improve the community," and pursued biomedical engineering because she wanted "a more direct connection with helping people." Her experiences as an educator are equally broad in that she has worked with young children, high school students, undergraduates and even graduate students and faculty. Her service experiences are mapped directly onto her teaching and research, and her lifelong learning reflections map school, hobbies, and self-care onto the themes of connecting with other people and seeing different perspectives that "let different areas of my brain work."

Jazmin's self-authorship is most evident in her discussion of lifelong learning, in which she draws on the image of clay on a potter's wheel to describe first feeling like the clay itself, molded by her experiences, and then like the potter, choosing how to shape the clay of her life. This shaping is particularly evident in her career journey as she describes moving from an engineering major that, for her, did not include significant collaboration or include research that directly impacted individuals, into a doctoral degree in biomedical engineering to pursue that impact, and then into a postdoctoral position where she finally believes she has "reached a level of research that will benefit people." She explains that her interdisciplinary learning experiences played a key role in that shift: "the liberal arts gave me a sense of creating my own masterpiece and I saw that I was not limited to electrical engineering." Thus in assembling her many and varied learning experiences, Jazmin constructs a professional identity in which she is increasingly "in control of shaping [her] future" and in which she "can re-form [her] career into a new work of art that may not be as I originally intended but is still a masterpiece."

Lance

Lance's Home page includes a picture of himself with his wife and two children, and he later explicitly talks about family and parenting in the context of both teaching and lifelong learning. His portfolio includes a personal statement (written for an external organization) that describes a holistic vision driven by contributing to the larger social good. This vision drives his research (on energy infrastructure), teaching (helping students to both reach their own goals and be of service to their communities), and service (coordinating activities to help community members and leading a Boy Scout troop). That is, he not only brings multiple dimensions of his life into his portfolio, but he also highlights the connections among them and explicitly threads them together into a narrative built on a set of core values.

Not surprisingly, a central thread in Lance's portfolio is the intersection of broad vision and focused purpose. He attends equally to his roles in teaching, research, service, and lifelong learning, and undergirds this balance with a consistent "big picture" orientation toward social good. For instance, although he narrows his motivation for becoming an engineer down to two reasons, these reasons are actually broad principles: to improve energy efficiency in the United States and to promote community education and action. These same principles permeate all his endeavors. As a teacher, he seeks to help students contribute to their communities, and he starts by building on what they already know. Although his portfolio's teaching reflection focuses only on homework assignments, he situates these within his larger intent to produce independent learners who can in turn contribute to their own communities. His research is characterized by both humility and influence when he notes that his contributions as part of a research team "were not earth-shattering in technological sophistication, but they had significant impact." His service activities are similarly meaningful; in teaching English to Spanish speakers, for example, he concretely identifies outcomes for improving both their employment situations and their ability to engage with their children's lives.

In the context of self-authorship, then, Lance's portfolio, though briefer than several others, is also one of the most multidimensional and perhaps the most integrated. He constructs a clear vision of himself and his values and articulates the ways those values shape the various dimensions of his identity. In part, this sense of integration may result from the inclusion of his personal statement as an artifact because the genre itself asks applicants to provide an integrated vision of their research, teaching, and service. At the same time, integration also emerges in his reflections on research, teaching, and lifelong learning, suggesting that his capacity for self-authorship reaches beyond that single application requirement. He presents an image of himself in family, in community, in the classroom, and in the lab that forms a cohesive whole.

Lindy

Like Edwin, Lindy focuses her portfolio almost exclusively on her engineering work and includes comparatively little personal information. Her Home page shows her in a lab coat working with beakers and pipettes, and the other pictures she includes are a relatively neutral professional headshot, a photo of her working in a lab (albeit less formally dressed in flip-flops and a tank top), and a photo with her Engineers Without Borders (EWB) group. Her portfolio is dominated by engineering images—representations of data, engineered objects, and professional society logos. Her About page briefly mentions hobbies, but these hobbies do not recur elsewhere in the portfolio (e.g., as instances of lifelong learning or service). Instead, her discussions of service focus on engineering organizations such as EWB and ASEE, and her discussion of lifelong learning focuses on pursuit of new knowledge about engineering and teaching. Her understanding of engineering becomes most clear in her discussions of teaching, where she positions engineering work in the context of "global dilemmas of population growth, resource depletion, environmental degradation, and social inequity," and notes that engineers must "develop innovative and sustainable solutions both globally and locally … [that] balance the economic, environmental, social, and technical dimensions" of these challenges.

Although Lindy's portfolio might be considered narrow in its focus on engineering work, it is characterized by her consistent passion for promoting sustainability, and, linked to that, her passion for integrating sustainability into engineering education. Her work has thus moved from research on sustainability issues into research on teaching sustainability to undergraduates, and she has added coursework in public policy to provide her with a more holistic understanding of the global challenges she seeks to address. She understands that as an undergraduate student she "failed to recognize the importance of engineering for improving human welfare," and it was her experience with her college chapter of Engineers Without Borders (which she helped found) that helped transform her own vision and inspire her graduate teaching, research, and service. Broadly, EWB helped her understand the need for context and application in learning, but more specifically it helped to develop a global perspective on the challenges of sustainability.

As with Edwin, it is more difficult to characterize Lindy's portfolio in terms of self-authorship because she focuses primarily on her work as an engineer and engineering educator and chooses not to explore the intersections of those dimensions with other facets of her identity. She does bring an intentionality and a consistent focus to this work that echoes the sense of purpose Lance presents, which suggests her ability to thread together the core aspects of her professional life. Her sense of self also seems closely connected to a sense of community, however, and those linkages may point to a position closer to a crossroads than to full self-authorship. For example, she notes that confidence in her own abilities as a researcher resulted from presenting her undergraduate research to the scientific community. Similarly, her experiences of lifelong learning are linked to established educational certificates (e.g., in public policy) or professional networks in her field. However, given the limited nature of her reflections on these experiences, it is tenuous at best to characterize her position more fully.

Discussion

As noted earlier, these portfolios represent creative constructions of identity in two ways. First, the participants construct *professional identities* by drawing together and commenting on various experiences, values, and passions they select to represent their professional selves. Second, and equally important, they construct *an identity for the profession* of engineering itself, not only through explicit engineering philosophies, but also through the ways in which they describe their work and their values. Both sets of constructions represent creative acts of engineering identity work.

Professional Identities

As the narrative vignettes above demonstrate, these participants adopted different strategies for creatively presenting and integrating multiple dimensions of their identities, particularly in terms of how and how much they integrated their "nonengineering" selves. In some cases, these nonengineering dimensions were presented, but not commented upon, as in Edwin's photograph of his family or Frasier's c.v. noting his work establishing groups for Jewish adults in his communities. These dimensions are represented, whether in text or in image, and are thus salient to participants' discursive constructions of their public selves, but they are not narratively integrated into their professional philosophies or personal journeys through engineering.

In other cases, participants clearly foreground such personal dimensions. For example, Gloria's discussion of lifelong learning foregrounds her passion for French and includes not only a narrative about her experiences in learning the language in school, but also photographs of her time studying abroad in France. Her experiences become a representation of both her love of the language and her more general commitment to lifelong learning, but she does not verbally link them to her work as an engineer. Similarly, Corbin's passionate discussion of his work on the Israeli–Palestinian conflict forms the core of his service discussion in his portfolio. But while this work implicitly overlaps with his desire to serve humanity through his engineering work, he does not explicitly make those connections in his reflections. That is, he does not discursively position the two aspects of his identity (engineer and peace activist) as stemming from a single core. In these cases, personal dimensions of participants' identity sit alongside their roles as engineering researcher and educator, highlighted and illustrated, but not necessarily integrated.

Other participants bring in personal dimensions of their lives in ways that do explicitly intersect with their engineering selves. Isabella, for example, in describing her passion for contract bridge, talks directly about the ways bridge is like her computer science work—mathematical, dealing in probabilities, centered on optimization. Jazmin incorporates her academic explorations in a variety of liberal arts courses into the narrative about her development as an engineer; she credits these explorations with helping her move from electrical into biomedical engineering to better match her own interests and values. For Denver, the intersection is not between engineering and interest but engineering and race. His career reflection highlights both his desire for research collaborations with HBCUs and his own positionality as a role model; similarly, his service reflection describes the faculty/staff/student mixer he sponsored for minorities in STEM. In both cases, his racial identity is visibly part of his work as an engineer in a university setting. Cases like Isabella, Jazmin, and Denver illustrate ways in which participants explicitly draw attention to the ways in which different dimensions of their personal and professional identities intersect.

Finally, Lance's portfolio stands slightly apart from the others in that the integration of the various dimensions of his identity is the most explicit, as crafted in his personal statement. That is, not only does he interweave discussions of his engineering research with his service as a Boy Scout leader and his learning as a parent, but he also positions all of these as rooted in a core set of personal values that he presents as a holistic vision. Moreover, rather than keeping his various roles (as educator, as researcher, as learner) separate, he consistently foregrounds the ways insight from one role crosses boundaries to integrate with and inform other roles. His intersecting identities consistently cooperate and reinforce each other.

Through their portfolios, then, these participants create complex, multidimensional representations of their identities as engineering professionals. It is important that, in drawing attention to the variations in how participants creatively represent the multiple dimensions of their identity, we are not suggesting that integration across differing dimensions is necessarily "good" or that lack of integration corresponds to lack of self-authorship or creativity. Instead, we use these distinctions as a means to illustrate the ways in which these emerging engineering researchers and educators actively self-author themselves, creatively constructing their professional identities as they select, feature, and intertwine various dimensions of their identity.

Identity of the Profession

Equally important, these narratives illustrate the ways in which reflective portfolios afforded participants the opportunity to craft discursive identities that placed engineering in the context of a larger set of goals, activities, passions, and people. Just as participants create themselves, that is, they create their profession. These portfolios are notable for the ways in which the participants consistently name the stereotypical external definitions of "engineering as the application of math and science" and "engineers as problem solvers," and then subvert, extend, or reshape those definitions by exploring both the contexts and the ends of such work. This act, we argue, is a creative gesture of self-authorship as the participants construct engineering for themselves rather than rest in received narratives.

In some cases, this creative reconstruction is explicitly delineated. Jazmin, for example, begins her engineering philosophy by explaining that "in simplest terms, engineering is problem solving. And to add a little more specificity, problem solving within a given set of constraints." She then adds context by noting the kinds of problems engineers solve and describing them as people who "are creative and use science and math to produce solutions to real-world problems." She pushes her definition further by asking, "But why do people become engineers?" to which she answers "to help others and create a better society," thus inscribing engineering with both a context and a purpose that supersede conventional stereotypes.

Other participants recreate the definition of engineering in ways that are more implicit; rather than discursively tracing the evolution from problem solving to service, they more directly represent their definitions of engineering as "advancing the human condition" (Denver) or "dream advancement" (Isabella). Through such phrases, these participants take agency in defining engineering in their own terms, oriented toward the values they seek to embody in their own work and share with their students and colleagues. They describe in detail their commitment to addressing challenges in energy, climate change, sustainability, human health and welfare, resource management, safety, and more. Equally important, they describe details of their own research, teaching, and service that actively and consistently address these challenges. Their constructions of engineering, in their philosophies and in the work they describe, all creatively subvert the stereotypical monolithic engineering identity embodied in characters from *Dilbert* and *The Big Bang Theory*. Rather than disconnecting engineering from both context and purpose, these participants' portfolios repeatedly create an identity for the profession that is anchored in real needs and oriented toward meaningful service. These created identities for engineering are, like participants' personal identities, critical acts of self-authorship as the participants move beyond the received definitions and stereotypes to reconstruct engineering in ways that match their own self-images.

At the same time, these portfolios creatively redefine not only *what* an engineer is, but *who* engineers are. In some cases, such as Denver's, this redefinition is explicit and intentional; as an African-American male, he sees himself as a role model and actively embraces activities that help redefine the demographic boundaries of the field. In other cases, the redefinition is implicit as participants include pictures of themselves with families or on travels, describe learning Spanish or cooking, highlight service to community gardens or youth organizations, articulate visions for social justice, and more. Both visually, by presenting images of themselves within and beyond engineering workplaces, and textually, by describing learning and service experiences beyond engineering practices, these participants have recreated the identity of "engineer" writ broadly to bring in a fuller range of interests, demographics, passions, and commitments.

Finally, these participant portfolios also creatively reconstruct the identities of engineers as (future) faculty members, specifically as they describe the integration of research, teaching, and service. The structure of the P2P project required students to illustrate and reflect on each of these components of their work, but it is the participants themselves who creatively bring them together and afford often equal weight to each. Where popular stereotypes see university faculty as "only" teachers who spend a few hours a week in a classroom or, more often, ivory-tower researchers who care only for their own lists of publications, these participants draw vital connections between the two, as when Lindy describes bringing her commitment to sustainability into capstone design courses or Gloria describes using her own industry experience as the basis for helping students expand their capabilities in creative thinking. Even when, as noted above, participants do not explicitly identify links among the three domains, often the same commitment to what Jazmin calls "creating a better society" drives all three dimensions of participants' portfolios, and lifelong learning offers them consistent opportunities to develop their skills and expertise.

These participants, then, create an identity—or more precisely, a set of identities—for engineering that includes more than simply the application of math and science, and for engineering faculty that encompasses more than research machines or lazy teachers. They are linked by a shared thread of humanitarian interest, one that is informed by their own learning as well as their own demographic and social identities. Their portfolios thus discursively create their profession in ways that broaden both what it means to be an engineer and who can become one.

Conclusions

The interpretive narratives that we have presented here demonstrate the power of portfolios as tools that enable participants to create holistic identities for both themselves and for engineering itself. The discursive acts that students undertake when they pair evidence of their achievements with narrative reflections serve as sites of identity construction that allow students the time and the framework to intentionally construct their career trajectories, their personal development, and the various facets of their professional roles. In doing so, as the portfolios in this study suggest, they are often able to make meaning for themselves in ways that enable them to negotiate among multiple dimensions of their identity. While the P2P project offered participants categories to help structure the portfolio, and the peer interactions provided a source of support and discussion, the reflective work of creative self-authorship belonged to the participants themselves.

As a result, these portfolios represent creative ways of thinking not about engineering products, but rather about the engineering field itself. The P2P project, as well as similar portfolio efforts, prompts students to be reflective practitioners rather than take for granted either their individual career trajectories or their field as a whole. Constructing a portfolio is a creative act of self-authorship that allows students to integrate their many intersecting identities and craft wholly new and unique ways of being an engineer.

Acknowledgement This material is based upon work supported by the National Science Foundation under Grant No. 1055595. Any opinions, findings, and conclusions or recommendations expressed in this material are those of the author(s) and do not necessarily reflect the views of the National Science Foundation.

The authors would also like to thank the researchers who have contributed to our coding: Wende Garrison, Rachel Kajfez, Martina Svyantek, and Ashish Agrawal.

References

Abes, E., & Jones, S. R. (2004). Meaning-making capacity and the dynamics of lesbian college students' multiple dimensions of identity. *Journal of College Student Development, 45*(6), 612–632.

Abes, E., Jones, S. R., & McEwen, M. K. (2007). Reconceptualizing the model of multiple dimensions of identity: The role of meaning-making capacity in the construction of multiple identities. *Journal of College Student Development, 48*(1), 1–22.

Bamberg, M., De Fina, A., & Schiffrin, D. (2011). Discourse and identity construction. In S. J. Schwartz, K. Luyckx, & V. L. Vignoles (Eds.), *Handbook of identity theory and research* (pp. 177–199). New York: Springer.

Baxter Magolda, M. B. (1992). *Knowing and reasoning in college: Gender-related patterns in students' intellectual development*. San Francisco: Jossey-Bass.

Baxter Magolda, M. B. (2004). Learning partnerships model: A framework for promoting self-authorship. In M. B. Baxter Magolda & P. M. King (Eds.), *Learning partnerships: Theories and models of practice to educate for self-authorship* (pp. 37–62). Sterling, VA: Stylus.

Baxter Magolda, M. B., & King, P. (2004). *Learning partnerships: Theory and models of practice to educate for self-authorship*. Sterling, VA: Stylus.

Capobianco, J. (2006). Undergraduate women engineering their professional identities. *Journal of Women and Minorities in Science and Engineering, 12*, 95–117.

Crede, E., & Borrego, M. (2012). Learning in graduate engineering research groups of various sizes. *Journal of Engineering Education, 101*(3), 565.

Crede, E., Borrego, M., & McNair, L. D. (2010). Application of community of practice theory to the preparation of engineering graduate students for faculty careers. *Advances in Engineering Education, 2*(2), 1–22.

Foor, C. E., Walden, S. E., & Trytten, D. A. (2007). "I Wish that I Belonged More in this Whole Engineering Group:" Achieving individual diversity. *Journal of Engineering Education, 96*(2), 103–115.

Jones, S. (2000). A conceptual model of multiple dimensions of identity. *Journal of College Student Development, 41*(4), 405.

McNair, L. D., & Garrison, W. (2013, June 26). *Raze the Silos: Using Digital Portfolios to Increase Integrative Thinking.* Paper presented at the ASEE Annual Conference, Atlanta, GA.

Patton, M. Q. (2002). *Qualitative evaluation and research methods* (2nd ed.). Thousand Oaks, CA: Sage.

Pierrakos, O., Beam, T. K., Constantz, J., Johri, A., & Anderson, R. (2009, October 18–21). *On the development of a professional identity: Engineering persisters vs engineering switchers.* Paper presented at the Frontiers in Education Conference, 2009. FIE '09. 39th IEEE.

Svyantek, M. V., Kajfez, R. L., & McNair, L. D. (2015). Teaching vs. research: An approach to understanding graduate students' roles through ePortfolio reflection. *International Journal of ePortfolio, 5*(2), 135–154.

Tate, E. D., & Linn, M. C. (2005). How does identity shape the experiences of women of color engineering students? *Journal of Science Education and Technology, 14*(5/6), 483–493.

Tonso, K. L. (2007). *On the outskirts of engineering: Learning identity, gender, and power via engineering practice.* Rotterdam: Sense Publishers.

Tonso, K. L. (2014). Engineering identity. In A. Johri & B. M. Olds (Eds.), *Cambridge handbook of engineering education research.* Cambridge, UK: Cambridge University Press.

Turns, J., & Lappenbusch, S. (2006, June). *Tracing student development during construction of engineering professional portfolios.* Paper presented at the American Society for Engineering Education Annual Conference and Exhibition, Chicago, IL.

Uneasy Stories: Critical Reflection Narratives in Engineering Education

Gillian Epstein and Yevgeniya V. Zastavker

Defining the Problem: Reflective Practices in Engineering Education

In the past decade, governmental, academic, and industrial leaders have been increasingly calling for critical changes in engineering education (NAE, 2004, 2005). To respond to this call, engineering programs throughout the country refocused their mission and aspiration to include explicit statements about supporting development of engineers who are not only experts in their fields but also possess critical thinking and related skills (see Example Mission Statements in the References section). Unfortunately, among other challenges, prevalent engineering pedagogical practices continue focusing on traditional delivery methods and suffer from content overload, significantly diminishing student opportunities to acquire and practice critical thinking (Andrews & Patil, 2007; Donawa, Martin, & White, 2007; Mina, Omidvar, & Knott, 2003; Romkey, 2009).

Defined narrowly within discursive engineering frameworks, critical thinking denotes debugging, synthesizing across learning, leveraging mistakes to improve outcomes, etc. (Claris & Riley, 2012, 2013; Facione, 1990, 1998; Mina et al., 2015; Paul et al., 2003; Phillips & Green, 2011). In other words, engineering education frameworks, if present, favor purely cognitive skills at the expense of self-regulation and personal dispositions (Facione, 1990, 1998; Romkey, 2009) and usually omit emphasis on reflection, reflective practices (Dewey, 1910, 1933; Schön, 1983, 1987), adaptive expertise (Facione, 1990, 1998; Hicks, Bumbaco, & Douglas, 2014; Lin, Schwartz, & Bransford, 2007; Zastavker & Epstein, 2015), and reflexivity, the latter three serving as different conceptions of critical thinking (Claris & Riley, 2012, 2013).

G. Epstein (✉) • Y.V. Zastavker
Franklin W. Olin College of Engineering, Needham, MA, USA
e-mail: gillian.epstein@olin.edu; yevgeniya.zastavker@olin.edu

D. Bairaktarova, M. Eodice (eds.), *Creative Ways of Knowing in Engineering*,
DOI 10.1007/978-3-319-49352-7_7

> Imagine an undergraduate engineering persona who embodies the best of such critical thinking and let's consider this student's limits: Focused Frank is not sure why he is doing engineering beyond the satisfaction of doing a task well; he does not often get to work with others; he works in a vacuum; he depends on himself to overcome any intellectual obstacles; he separates his personal life and values from his academic work and does not see synergistic connection between them; the non-engineers in his life do not understand the impact of the work he does; when he imagines his future career, Focused Frank does so in a purely technical framework that does not consider values, politics and context, all of which circulate in a "separate" world. When Focused Frank completes his undergraduate education, he may excel technically but his work will likely be split off from his personal values and goals and, by extension, will be disconnected from the context and community in which he lives. This persona has not been equipped to leverage technical insights and understandings to better the world, to solve compelling needs in our contemporary world (NAE, 2004, 2005).

The knee-jerk solution to helping Frank before he graduates is to provide him with focused modules or a course on developing critically reflective and adaptive skills necessary to integrate his work with his personal values and ambitions. However, to accomplish this we ask faculty to create critical reflection experiences for students. And here we hit upon a quiet problem in academia: we presume faculty are equipped, able, and willing to promote and teach critical reflection within the classroom and beyond. In fact, in our preliminary work as substantiated by literature, we have many Focused Franks on the faculty who, regardless of whether the development of critical thinking is an explicit institutional goal, may be unable to articulate what critical thinking means to them (Ahern, O'Connor, McRuairc, McNamara, & O'Donnell, 2012; Lin et al., 2007; Paul, 2004; Phillips & Green, 2011). Other faculty may hold deep convictions about curricular emphasis on critical reflection, yet may lack theoretical frameworks and pedagogical expertise to teach it. If we can even agree that our student persona deserves an opportunity to know himself better and in so doing make a more connected impact on the world, we must acknowledge that despite our best ideological intentions, we as faculty do not know how to support development of critically reflective students (Ahern et al., 2012; Hicks et al., 2014).

Moreover, if we privilege self-knowledge as a mode for students to become critically reflective, connected citizens, then we must consider the entirety of student experience when we approach the question of how to support students' development as critically reflective citizens. Notably, in academic solutions for "teaching" critical reflection, the ghettoization of student life and postgraduate planning from purely academic endeavors remains unchallenged (Wilson et al., 2014). To educate a student who integrates life and work, we must integrate development of critical reflection throughout the learning continuum in a way that does not cordon off academic learning from learning outside of the classroom. When we do consider the diversity of student life and experience that exists in a student body, our responsibility to teach critical reflection increases because students may overlook their unique perspectives and motivations. By attempting to fit into the dominant paradigm of engineering talk, look, and practice, students may not recognize their personal context and may thereby disconnect from learning engineering. In comparison, engagement in critical reflection with their diverse life experiences could help students emerge as new voices and leaders who leverage their technical understandings to benefit their communities in new ways (Wilson et al., 2014). Leveraged well, critical reflection may fuel students' motivations, which, according to self-determination theory, just may allow integration of their technical work with a sense of autonomy, relatedness, and competence that inspires and enables them to do good in the world (Deci & Ryan, 2002; Hicks et al., 2014; Ryan & Deci, 2000; Shahar, Henrich, Blatt, Ryan, & Little, 2003; Turns, Sattler, Yasuhara, Borgford-Parnell, & Atman, 2014; Van den Broeck et al., 2010; Wilson et al., 2014). Failure to teach critical reflection becomes a liability for any engineering education program hoping to nurture diversity.

What We Know and What We Need to Learn

There is a great deal of research in critical reflection, critical thinking, reflective practice, reflexivity, and adaptive practice (Claris & Riley, 2012, 2013; Fook & Gardner, 2007; Garnder, 2014; Hicks et al., 2014; Johns, 2005; Mina et al., 2003). Much of it is shaped by education literature stemming from such diverse fields as psychology, sociology, linguistics, nursing, etc. Of note, however, is that STEM education and specifically engineering education have not embraced critical reflection as a field of focused study. For example, although reflection has been gaining wider academic discourse in engineering education, only a small number of publications make reflection a primary focus (Sepp et al., 2015). To our knowledge, in the last ten years, *Journal of Engineering Education*, a premier publication in the field, printed only one manuscript centered on reflection (McKenna, Yalvac, & Light, 2009).

Engineering education literature on reflection, critical reflection, critical thinking, reflective and adaptive practices is presented in the contexts of problem- or project-based learning (Chang & Wang, 2011; Choulier, Picard, & Weite, 2007; Cloete, 2001; Khisty & Khisty, 1992; Lombardo, 2004; Mina et al., 2003; Papadopoulos, Rahman, & Bostwick, 2006; Polanco, Calderon, & Delgado, 2001; Vos & de Graaff, 2004), design (Adams, Turns, & Atman, 2003, Khisty & Khisty, 1992; McKenna, 2007; Turns, 1997; Walker, Cordray, King, & Brophy, 2006), experimental work (Bruno & Anderson, 2005; Miller & Olds, 1994), research (Doel, 2009; Hunter et al. 2007), and ethical considerations (Catalano, 1993; Chang & Wang, 2011; Rayne et al., 2006; Swalie & Kreppel, 2001; Wolverton & Wolverton, 2003). Operationalization of these practices in engineering education is performed in a variety of ways ranging from embedding writing assignments into STEM content-driven coursework (e.g., essays, portfolios), inclusion of surveys and class discussions, and introduction of stand-alone courses focusing on critical thinking (for examples, see Sepp et al., 2015 and Hicks et al., 2014). However, theorization of reflection and related concepts in engineering education seems to be limited primarily to cognitive approaches.

By far, the Delphi report, "a statement of expert consensus for purposes of education assessment and instruction" developed by a group of leading academics in the field, seems to have most significantly captured the imagination of engineering educators (Facione, 1990, 1998). Unfortunately, the two-faceted approach to critical thinking (and its translation to reflection) described in the report, that of cognitive skills and affective dispositions, is mostly operationalized in engineering education as that of cognitive skill development, such as analysis, evaluation, inference, interpretation, explanation, and (rarely) self-regulation. Affective dispositions of critical thinking, such as fair-mindedness, honesty in facing biases and prejudices, self-confidence, and others have been left unexplored and un-operationalized (e.g., Hicks et al., 2014). The Delphi report was later popularized by Richard Paul in a small pamphlet called *Miniature Guide to Critical Thinking: Concepts and Tools* (Paul, 2008), which served to further promote cognitive approaches to critical thinking and reflection over other approaches. Claris and Riley (2012, 2013) hypothesize that this may be due to "the ubiquity of the phrase *critical thinking skills* and a lack of *critical* attention to its usage in echoing the accepted jargon" (Claris & Riley, 2012, 2013; accent added by the authors of this work). The comfort of this engineering education paradigm with positivist ideology that values rigor, objectivity, and detachment created ideal conditions for cognitive approaches to continue to dominate discourse about reflection and critical thinking (Rolfe, 2014; Schön, 1987).

Interestingly, it is this ideology (or more specifically, technical rationality, the positivist epistemology of practice) that was radically critiqued by Donald Shön, the architect of the reflective practice perspective. Reflective practice is based on the premise that tacit knowledge is embedded in practice and is generated through the process of reflecting on and in practice (or action), which is "at least equal value to knowledge derived by academics from empirical research" (Fook, 2013; Rolfe, 2002; Shön, 1987). Engineering education seemed to have resonated with this theoretical approach to reflection; it was argued that reflection on and in action may be used as a way of approaching "wicked problems," the open-ended, ill-defined problems that cannot be simply solved using a technical rational approach based on theory and best empirical evidence (Hicks et al., 2014; Rittel & Webber, 1973; Rolfe, 2014). In fact, the recent movement towards student-centered and activity-based approaches to teaching and learning engineering with open-ended ill-defined problems at their core seemed to have offered an opportunity to use reflective practice as a way of learning; however, these teaching approaches also seemed to have assumed that reflection implicitly teaches itself given the experiential nature of teaching (Zastavker & Epstein, 2015). This may be an additional reason that to date the engineering education landscape doesn't seem to embrace reflective practice "in practice" and the results of studies about the efficacy of reflective practice in our classrooms remain opaque, inconsistent, and insignificant (e.g., Claris & Riley, 2012; Hicks et al., 2014; Hickson, 2011; Mina et al., 2015).

In recent years, critical perspectives have been used to theorize engineering education practices and engineering education as a field (Claris & Riley, 2012, 2013; Riley, 2003, 2008, 2015; Riley & Claris, 2006). Perhaps, Claris and Riley (2013) capture it best when they say,

> We long ago reached a certain "situation critical" in engineering, in which engineers may have abundant logical thinking skills but no practice of thinking critically about problem framing, power relations within the profession, hegemonic epistemologies of the discipline,

> or reproductive practices of engineering education. This leaves society with technologies that replicate power relations of the status quo, the profession with intractable problems around underrepresentation of white women and women and men of color, and few tools for understanding or talking about these problems, let alone acting effectively for change. (Claris & Riley, 2013)

In other words, if we are to prepare a next generation of engineers capable of tackling "wicked" problems that are deeply embedded into social and cultural environments, engineering education must use critical perspectives to help students create a vision for and operationalize the ways in which engineering may support social changes. Critical perspectives therefore are an integral constituent in the transformative value of critical reflection, something that is yet to be brought to bear in the way we currently educate our engineering workforce of tomorrow (Claris & Riley, 2012, 2013; Fook, 2013).

In the past decade, a few authors began addressing the value of relational approaches in engineering students' development vis-à-vis reflection as a way of learning. Walther et al. (2007) argue that "a *holistic integrative learning* process across all domains of the students' learning could be supported by explicit and directed reflection" (Walther et al., 2007). They propose that, in addition to the cognitive approaches to reflection, i.e., description, interpretation, and decision-making, students are engaged in unpacking affective aspects of their experiences. In fact, they maintain that emotional triggers may be used to facilitate reflective activities with students (Sepp et al., 2015; Walther et al., 2007, 2009, 2011). Unfortunately, this type of argument is an exception rather than a rule in dominant engineering education paradigms.

With the establishment of the University of Washington's Consortium to Promote Reflection in Engineering Education (CPREE) funded by Helmsley Charitable Trust, engineering education as a scholarly field has found a more focused effort towards the understanding and shaping of critical reflection paradigms and practice; however, this effort is still a neonate (CPREE, 2014). The CPREE effort is centered around identification and development of materials that support student learning through reflective practice. As such, even this unique for engineering education endeavor focuses minimally on holistic development of engineering students through reflection (e.g., through deeply reflexive, personal, and affective development that integrates both personal and professional aspects of their educational experience) and does not focus on understanding faculty beliefs and values relevant to reflective practices or development of faculty and institutional culture around reflection (CPREE, 2014).

Our preliminary work within Olin College[1] (Zastavker & Epstein, 2015) indicates the curricular activities, Passionate Pursuits, Olin Self-Study, Independent

[1] Olin College of Engineering (www.olin.edu) was established in 1997 as a response to multiple calls by governmental, industrial, and political leaders to revolutionize engineering education. Drawing upon the best pedagogical and curricular practices, Olin features a range of teaching methods ranging from more traditional to nontraditional, including project- and problem-based learning, do-learn and hands-on environments, teamwork, and self-directed learning. With a belief that learning happens in and outside of a classroom, Olin's learning continuum allows students to

Study activities, research, and clubs. From the College's inception, its admission, academic, and student life practices have focused on creating and maintaining a gender-balanced student body. To date, Olin has graduated 11 classes of students who continued on to graduate schools, earned generous scholarship, won fellowship opportunities, started their own businesses, or pursued careers in existing companies. The critical reflection process requires [individuals] to use all of themselves, to take into account the emotional, social, mental, physical, and spiritual. The process is one that includes recognizing and working with emotions and thoughts, recognizing the influence of social context and the physical world and the impact of what is meaningful (Gardner, 2011, pp. 70–71).

This is a holistic approach that allows all of oneself to be involved in the practice of "being" in the reflection space and "doing" reflection. The "being" aspect of critical reflection allows for creation of an inner space where a reflective dialogue about personal self-knowledge that is subjective, tangible, authentic, and existential can happen (Carper, 1978; Garnder, 2014). The "doing" aspect of critical reflection has to do with the transformative and purposeful nature of reflective processes that allow one to change the viewing lens on self and context, opening opportunities for creation of hypotheses, testing those hypotheses, analyzing the results and trying it all over again, all of it done here and now in professional practice and practice of selfhood development (Rolfe, 2014).

Donald Schön (1983) called this kind of "doing" reflection-in-action: reworking one's narrative, developing one's new story, designing one's new path, engineering a new identity through a different set of talking points and critical lenses, etc. However, in the world of engineering education, the terminology of reflection-in-action has been encoded into a much more discursively acceptable "learning by doing." Unfortunately, by giving such a physical embodiment to a reflective process, this metaphor divorces the hard work of critical reflection, the one that integrates emotional, social, mental, physical, and spiritual, from the engineering practice. This implies that "thinking about doing something while doing it" (Schön, 1983,

---

pursue their passions through co- and extra-following trends: faculty and staff do not have theoretical frameworks that scaffold understanding of critical reflection; culturally, only a few faculty and staff embrace critical reflection in their learning environments but still share concern about not having pedagogical tools, methods, and expertise to support student development. Notably, these are trends at an institution whose explicit mission is to produce graduates capable of solving Grand Challenges of Engineering and whose implicit aspirations are focused on leveraging critical reflection to afford development of such graduates: an agile institution, which boasts student-centered curriculum and holds students' intrinsic motivations at the core of its practice (Olin 2016). Yet Olin faculty and culture suffer from lack of understanding and ability to practice student development of critical reflection skills. So, what is the landscape of critical reflection in other institutions and what are the unique features of the programs that are particularly successful in development of critically reflective students? What role do the nonacademic programs within higher education play in supporting development of critical reflection skills? What epistemic beliefs about critical reflection are held by those who create students' experiences in higher education? And how do those epistemic beliefs shape culture and operationalization of critical reflective practices on the campuses throughout the USA? Our approach to critical reflection is best summarized by the work of Fiona Gardner (2011) when she says,

p. 54) is no longer about solving a wicked problem of building engineering selfhood; rather, it is now about "building" an artifact evacuated of the meaning that artifact may hold beyond its physical representation.

And here we run into the second quiet problem in engineering education. We presume to prepare students to be the generation that understands systemic opportunities and challenges affecting global quality of life, the generation that relates to the world around them in holistic ways, the generation that "solves the grand challenges facing society in this century" (NAE, 2009). However, we are missing a critical step, the one that provides students with tools necessary to relate to the world around them. To understand the ambiguity of the open-ended problems that include both objects and people, to be able to relate to the discrepancy between oneself and others, and to be willing to accept uncertainty of reality that is not predetermined by rules, categorizations, and stereotypes, one needs to create an authentic personal relationship between oneself and others, and between oneself and challenges to be solved. In the words of Barbara Carper, designer of the theoretical framework describing four fundamental patterns of knowing in nursing, "the individual is considered as an integrated, open system [between oneself and their challenges] incorporating movement toward growth and fulfillment of human potential… [and] the recognition that each person is not a fixed entity, but constantly engaged in the process of becoming" (Carper, 1978). This is the core of the personal pattern of knowing, one of the four patterns of knowing in the Carper's framework, as described in Table 1 below.

This idea of "knowing" oneself to "know" another is also resonant with the postulate put forward by Blatt and colleagues about centrality of interpersonal relatedness and self-definitions in the developmental process throughout one's life span (Blatt, 1974, 1995, 1998; Blatt & Blass, 1996; Blatt & Shichman, 1983; Blatt & Zuroff, 1992 in Shahar et al., 2003). Relatedness, one of the basic psychological needs underlying individuals' motivations and behavior and supporting well-being (Deci & Ryan, 2002; Ryan & Deci, 2000), refers to people's fundamental need to be connected to others, to belong to a group, to establish long-lasting loving and caring relationships with others, etc. Self-definition refers to the individuals' need to estab-

Table 1 Carper's ways of knowing in nursing framework (Carper, 1978)

Way of knowing	Description
Empirical	A way of knowing that involves scientific evidence that can be empirically verified
Personal	A way of knowing that engages practitioner's self-knowledge, mental models, attitudes, behaviors, and feelings
Ethical	A way of knowing or behaving that involves moral questioning, inclusion of ethical frameworks, and decision-making processes that require the right moral action or nonaction
Aesthetic	A way of knowing that includes the art of practicing involving sensing, perceiving, and feeling, and responding to practical situations within the moment

lish a stable, coherent, authentic, and positive sense of self. According to Blatt et al., the two processes develop synergistically and their successful integration and negotiation supports individuals' optimal development (Blatt, 1995; Blatt & Blass, 1996; Blatt & Shichman, 1983 in Shahar et al., 2003). In other words, we cannot develop the next generation of students who will tackle the problems of others if we cannot develop students who will tackle the problems associated with knowing oneself.

If engineering education is to shift towards a more contextual, humanistic paradigm that is deeply imbedded in practice of solving human problems for the betterment of human lives, we need to imbed ourselves "in the kind of knowing involved in reflection (a subjective, integrated, and inferential kind of knowing)" (Turns et al., 2014).

Critical Reflection: Now What?

Our emergent process of critical reflection is grounded in the idea of the reflexive connection between identity and narrative. By interrogating self through narratives and stories, we believe we can bring together our personal and professional selves, we can position our personal multiplicities within our professional practice and reposition our professional identity within our personal subjectivities. Jerome Bruner (2002) argued that "it is through narratives that we create and re-create selfhood, that self is a product of our telling… if we lacked the capacity to make stories about ourselves, there would be no such thing as selfhood" (Bruner, 2002, pp. 85–86, in Alsup, 2013). In this way, we are working to narrate ourselves to our new identities and our new positioning vis-à-vis the world around us.

If we, as academic institutions, are going to teach critical reflection, we are asking critical reflection not to be a moral abstraction but rather a construct with a voice that we can scaffold and nurture so students learn to iterate on stories that (re)connect them to themselves and to the world. Selves are shifting, they are perspectival and unstable, they can involve agency and choice, be narrated in different ways that are all true, or be dissonant. Without interrogation of dominant narratives and stories, it is very easy, even at Olin, an iconoclastic edgy school, to become a part of dominant paradigms rather than fully embrace the unique diverse voices about who an engineer is and what her role is in the world. It is this process of interrogating, shaping, and narrating our emergent narratives of integrated personal and professional selfhood that we call "doing" when we speak about students' reflective work, as described below.

Design Thinking and Student Life: What Do They Have in Common?

At Olin, a school that prides itself on innovation in teaching and learning engineering, what we term the "design stream," a series of design courses that build upon one another, makes a big impact on student experience and school identity. Based on the idea that the purpose of engineering is to better people's lives, our students learn to practice engineering by starting and ending their thinking with a user in mind. All students take a "User-Oriented Collaborative Design" (UOCD) course their second year, and explore the process of understanding their user in every stage of the design process. Students learn to interview users, and most importantly to ask critical reflective questions about what the user tells them: e.g., when a student asks a sustainable farmer what he or she needs to make work go better, the farmer might request a specific tool or launch a specific idea. It is the student's job to question that assertion, perception, or need: does the farmer fully "know" what he or she wants? What is the context of the farmer's request? What problems might the farmer be missing? How do other perspectives change the picture of what the farmer needs? What questions can the student ask to get a better understanding of the farmer's needs?

If critical reflection is at the centerpiece of how students experience the design stream, our call is for additional or for different modes of reflection. Notably, in the design process, students are specifically taught to erase themselves from the user equation, meaning that they must take care not to impose their ideas, experience, or assumptions upon the subject whom they interview. So while students engage in a critically reflective process to solicit and analyze information from the user, they carefully refrain from including subjective bias in the process: students learn to reflect on the stories of others and to suppress their own stories, except within the framework of trying to inductively derive understandings that will make them better designers.

And we thereby construct an interesting irony in our learning community: we would never presume to have Olin students design for a user without a careful critically reflective process that actively wonders about the gap between what someone says or perceives, versus what they are not yet telling us, what they implicitly mean, and how these meanings circulate in a larger context and community. And yet we presume to ask Olin students to design for their own life without this same level, or perhaps even any level of reflective curiosity: after 4 years of learning to make their personal values, context, stories, and voices invisible in the work they do, students are suddenly asked to tell us (as well as prospective employers, funders, academic and scholarship committees) what they plan to do with their lives after college. For these very high stakes questions of how our students plan to make a difference in the world and how they want to engage with global challenges, we believe it is fundamentally important to show their voices the same respect and attention that our students are taught to show users. Because like any user, the student's self-knowledge operates in a complex world of implicit, explicit, contextual, dominant, nondominant, subversive, and uneasy understandings. Meaning that when a student tells us "I am going to be an engineer because I come from a family of engineers," or "I want to be an engineer so I can work on real problems" or "I don't know why I am in an engineering program and when I graduate I'm going to follow my real passion for x, y, or z" there should be space and scaffolding for follow-up questions and for reflection. With a deeper analytical understanding of what students say they want and need, we believe students become more connected with themselves and hence will be able to better connect with others in whatever career they might design, now or later.

"What's Your Story?": Engineering Critical Reflection at Olin

The genesis of our critical reflection courses, the 2014 offering entitled "Critical Reflective Writing" and the more recent 2016 iteration titled "What's Your Story? Narrative with Purpose," navigates the musings and critical reflections presented above. In both versions of the course, we have noted different themes of relatedness that reliably emerge. These themes uncover different experimental reflection modes in terms of how students relate to themselves, to others, to the thematic they are uncovering or struggling with, and how they coalesce these discoveries into storytelling voices. These themes range from discussions of race and ethnicity, to positioning one's disability in the world of engineering, to crafting one's own sense of gender identity within a practice that privileges masculinity. In what follows we unpack just one of these themes, that of gendered patterns of "being" in and "doing" engineering, a theme that is pregnant with hidden assumptions, embedded stereotypes, and implicit biases about engineering.

Notably, Olin has pursued gender balance in admissions, and has a laudable reputation as an engineering school where more or less half of the student body is made up of women (U.S. News and World Report, 2015). Also notably, like most institutions doing the right thing in terms of equity, the conversation about integration of women into engineering within our Olin culture is not a dominant one and/or chiefly takes place in dedicated forums, e.g., our popular "Gender and Engineering" cocurricular activity—an activity that students take for a nondegree credit and faculty teach as a volunteer effort; gender intervention during a required first-year design course before students move from individual to team work; Women Open Houses, etc. While singular conversations in classrooms and cocurricular activities fill an important need by sharing stories about being a woman in engineering, they happen outside of our "formal curriculum" and are not positioned to offer students the support to do rigorous, iterative analytical processing and writing to uncover their own stories. And forums like Olin's Women's Open Houses necessarily focus on "positive" stories, working to recruit girls into STEM and spinning visions where girls can picture themselves as engineers.

Our course thereby fills an unmet need in our community: it provides a space for "uneasy" stories—for stories that might not fit the Olin norm we narrate of gender equity and opportunity in engineering. And as the term implies, "uneasy" stories are not always easy to access: by providing the scaffolding, iteration, feedback, and support for the discovery and construction of these stories, our course offers a space for students to hear and voice self-understandings they may not have been able to include in other community reflections about gender in engineering.

From our student stories, we have anecdotally found that many female students who committed to 4 years of intensive undergraduate engineering education at Olin have significant unprocessed and unvoiced stories about engaging in a traditionally male-dominated pursuit. In other words, these students struggle—usually silently—with relatedness through the framework of gender (Shahar et al., 2003; Van den Broeck, Vansteenkiste, De Witte, Soenens, & Lens, 2010).

While it does not solve the challenges of relatedness through the prism of gender in the context of engineering, the critical reflective processing of these stories makes the challenges visible. Students can suddenly witness and watch the choices they make within the ideological framework of gender and engineering: how they navigate, negotiate, repress, exaggerate, physicalize, question, simplify, or own the schisms that gender creates in terms of feeling a sense of belonging within our classroom and college community. And with this visibility comes the possibility for curiosity, reflection, empowerment, and perhaps even change: what patterns do they uncover? What changes could they make? What assumptions do they have? What messaging are they disrupting? What ideologies are they bolstering? When our female student struggles authentically with the reality of navigating engineering, we increase her opportunity to craft unique, individualized, and purposeful pathways of relatedness that integrate, rather than compartmentalize, her experience.

"What's Your Story?": Emergent Questions of Gender and Identity

Significantly, none of the assignments for either iteration of our two critical reflective writing courses specifically asked any student to reflect on gendered subject positions within the context of engineering. Rather, by design we keep our prompts abstract, asking students to write what we term "micro stories"—small stories from the student's life in which she is the primary agent, but may not know why the story has stuck with her, has not necessarily processed or analyzed the importance of the story, and begins or deepens opportunities for analysis by composing and sharing the story. In that sense, for some students the theme of gender becomes a surprise that sneaks up on them—yet one that is imbricated in the very fabric of how they ask and answer intellectual questions, or display scientific or academic curiosity. In what follows, we unpack this process using a case study of students A. and C. A. started her journey in our course by writing about an exchange with her father during childhood:

> *Growing up, my family and I used to take walks together...The four of us piled into the car and drove to the park. We would get out and I would walk ahead with my dad while my sister and my mom followed. While we walked, my small quick excited strides and his longer more laborious ones, I would ask him questions. As I looked around, I would ask him the first thing that popped into my head. About why the sky was blue and why the grass was green. About what he did every day in his mystical cubicle. About how we thought and what it meant to be alive. He never grew tired or frustrated. He chose his words deliberately, making sure to bring topics down to my level, growing my curiosity while also satiating it.*
>
> *Though he'd never admitted it, my Dad always wanted a son. Not because he was patriarchal but because he felt he could share his passions better with a son. My [older] sister fell in love with Polly Pockets and Barbies early and quickly. She wouldn't have any of my Dad's toys. When I was born, my Dad was determined to treat me as he would a son. He made me just as comfortable with Legos as with the princess Barbie expansion pack. We would work on his car right before my ballet lessons. I think these questions made him proud. He saw in me a way to connect through transferring his knowledge.*

When she presented this segment of her story in class, A. processed and framed it as a happy memory about relatedness: she was narrating a positive relationship with her father, and how he nurtured her interest in exploring proto-engineering questions about how things work. Instructor and class feedback surprised her.

Using the analytical close reading methods we guided students to direct at any written piece, we questioned and pointed out patterns within the narrative, many of which had to do with gender: what does it imply when you and your father walk "ahead" of your mother and sister? Does this suggest a hierarchy of affinities and explorations that are appreciated in your family? If, like your sister, you had rejected your "Dad's toys," would you have been able to claim your father's company? If not, what does it mean to you that the conditions for your father's attention are that you "share his passions?" In this story, how do opportunities for stature attach to asking proto-scientific (why is the sky blue, why is the grass green) questions? Regarding this link between scientific questions and stature, note that the discussion with A.'s father is "growing" her curiosity, and her effort to keep up with his interests means she tries to match his "longer" strides. And finally, what sorts of "knowing" (explicit versus intuitive) about gender does this piece exhibit, as in the following claim: "Though he'd never admitted it, my Dad always wanted a son." And how does this declaration point to a pattern of both disclosing ("always wanted a son") and repressing ("never admitted it") the understanding that A. needs to show interest in proto-scientific questions and play (which in this piece are coded as "male") in order to claim her father's love? This pattern of uncovering yet occluding is literalized by A.'s immediate qualification about her assertion that her father always wanted a son—she writes that it is "not because he was patriarchal" but "because he felt he could share his passions better." Here A. anticipates and quickly dismisses the possibility that her father represents a patriarchy—a family or social structure where men literally and figuratively dominate. Unwilling to explore this possibility, A. transmutes questions of male control into a lexicon of "passions," meaning that her father's interests are seemingly separate from gendered issues of power, and yet, A. has shared with herself and with us the contradictory observation that her father engaged her in these "passions" *because* he resolved to treat her "as he would a son."

This narrative dance of contradictory observations about how and why A. has particular interests and questions gets more complex the closer we continue to look. For example, in her story A. poses questions to her father of her own volition, and exuberantly chooses to walk and talk with him. Yet she also positions herself as having little choice about this alignment with her father: "When I was born, my dad was determined to treat me as he would a son. He made me just as comfortable with Legos as with the princess Barbie expansion pack." Here, the father's determination and choice frames the relationship, and since it begins at birth A. appears to have very little self-regulation in terms of her affinities. And, even in the daughter's play the father operates as agent: "He *made me* just as comfortable with Legos." Who "made" A.'s interest in becoming a maker within the world of engineering? Does A. perform curiosity for (male coded) approval in the realm of engineering? To what extent does A. feel intrinsic curiosity? What are the questions she wants to ask for herself, as opposed to engage or please (a masculinized) someone else?

"What's Your Story?": A Case for Exploring Uneasy Stories

One might wonder why make the effort to critically reflect upon A.'s story? If she feels happy and positive about this fragment of an origin story regarding how she came to position herself as someone within the traditionally male gendered field of engineering, why make trouble? Why uncover patterns of contradiction and complexity in how and why A. asks questions and forms interests and alliances in this story? Why or how could that be an important mode of growth for A.? Because, we find, those complexities point to silent, un-held conversations that unexplored can coalesce as black holes in self-knowledge and hence in relatedness not only to coursework in engineering but also to the project of designing one's life and career. A.'s continued writing is a fascinating case in point. In a subsequent assignment, we asked students to compose a micro story about a significant moment in their "professional" life (meaning within the context of any academic work). Interestingly, although she makes absolutely no explicit connections to her initial personal micro story and was in no way asked to make such connections, in the following excerpt the complex questions about A.'s motivations and affinities erupt as unavoidable and seismic gaps in self-knowledge and relatedness:

> *I distinctly remember one day a couple of us were talking with J. [engineering professor] about our passions and our dreams for the future and how these ideas related to our current education and what we were doing at Olin for research as well. I remember saying that at that moment I felt as though my passion was education; not just researching it in engineering teams, but also being hands on with students. I wanted to invoke passion in children and watch them grow into beautiful people. J. quietly listened to me and seemed to be thinking deeply. After I had finished, I remember he asked me, "A., why did you come to Olin?" I remember not thinking too much about it and said "I don't really know." This question has sat with me for quite a while and I don't feel as though I have a true answer. I know that it wasn't a very thought out decision. I know that I had a very different idea about what I wanted to do with the rest of my life than what I believe today. This question has sat with me for so long and I'm still not sure I know the answer to it.*

Here, A.'s relatedness to the project of engineering manifests as feeling lost as opposed to feeling connected: in response to the question of why she came to Olin for an engineering education A. replies "I don't really know." The subsequent paragraph of the excerpt loops through this idea of not knowing, inscribing feelings of inefficacy at analyzing and uncovering insight about her own motivations and directions: "This question has sat with me"; "I don't feel as though I have a true answer"; "I know it wasn't a very thought out decision"; "I know that I had a very different idea about what I wanted to do with the rest of my life"; "This question has sat with me for so long and I'm still not sure I know the answer to it." Again using the lens of analytical attention that literary criticism and critical discourse analysis offer, we see A.'s language reinscribe her conceptual idea of not knowing: her language repeats as opposed to uncovering new ideas, as evidenced by recycling phrases like "this questions has sat with me" and "I don't feel as though I have a true answer." Similarly, the only thing A. does "know" within the structure of these assertions is that she does not have a firm grip on self-understanding or her own direction: "I *know* that it wasn't…very thought out"; "I *know* that I had a very different idea…

than what I believe today." And finally, A.'s metaphorical self-representation as an immobilized subject contrasts sharply with her personal story's tale of a girl with curiosity and intentionality. When asking questions to please her father, A. makes "small quick excited strides" alongside her father's "longer" ones: she is on the move, trying to match her father's gait in contradistinction to "following" behind like her mother and sister. Whereas when asking questions directed at her own motivations and intentions, she associatively feels indefinitely stagnant and stationary: "This question *has sat with me for so long* and *I'm still not sure* how to answer it." In other words, have A. walk in her father's shoes and she literally follows in her father's footsteps and heads right for engineering school. Ask A. to walk in her own shoes, and suddenly she doesn't know where she's going. But she is not merely lost—she is indefinitely seated, metaphorically going nowhere.

Within the context of a writing assignment or writing class one might view A.'s excerpted piece as something she needs to "fix": voicing unresolved contradictory ideas, cycling repeatedly through the same phrasing and concepts, and expressing an immobilized lack of comprehension or vision are not often considered desirable effects in writing intended to engage oneself and one's audience. In the context of our critical reflection course, we instead consider these narrative "mistakes" as rich opportunities for curiosity, deeper thinking, and deeper learning. To access this depth, in addition to directing literary techniques for tracking language, content, and structure to pose and explore meaningful questions, we also model and practice other modes of listening and investigating. Our students use a "step back" (Keegan, 2002) model when presenting their pieces, a structured process where they share the piece, answer clarifying questions, and then listen and take notes on analytical discussion that a partner or larger group holds about the piece. With this process, the student who presents gets to take in multiple points of view on the potential meanings of her micro story; conversely, the students who offer feedback hone the ability to ask significant clarifying questions and discover analytical questions and interpretations—skills we then ask them to direct at questioning the meanings of their own micro stories.

"What's Your Story?": Increasing Relatedness Through Elderly Partnerships

To further develop student abilities to ask analytical questions, we provided them the chance to partner with an elderly person from a local Council on Aging; students inductively derived meaningful themes from their first two micro stories, and translated these themes into an interview protocol with five specific questions (and associated thematic "constructs") for their elderly partner (Jacob, 2012; Takacs, 2003). This exercise aimed to accomplish multiple objectives within the framework of relatedness: bridge the generational gap exaggerated by the undergraduate residential college experience, where students primarily live, work, and socialize with peers; give students perspective on the themes emergent in their micro stories by exploring these same themes with a relative stranger who is at a very different stage of life; encourage students to ask more and better questions about important themes they uncover; and to experience the relative effect of their question in the real-time connections made during the interview.

To scaffold this interview opportunity, we had students derive questions from their first "personal" micro story and their first "professional" micro story. In keeping with our objective of promoting more integration between personal and professional spheres, we then had students brainstorm connections between the two sets of questions. (Note that the boundary between "personal" and "professional" can often be quite artificial, something that students come to discover as they advance through the course but which may not be a comfortable or generative idea at the outset). From this bridging students derived a new set of interview questions that integrated personal and professional themes from their work.

With this scaffolding, A. first designed "professional" and "personal" interview questions that emerged from her own story and then made intentional connections between emergent questions about gender in her personal micro story and emergent questions about motive and professional direction in her professional micro story:

> ***Professional Question***
> *What have you viewed as "traditional"/"typical" paths expected in your professional life? Have you stuck to these or changed your path? (**Constructs**: expectation, tradition, society)*
> ***Personal Question***
> *How has gender played a role in your life? (**Constructs**: gender, society, cultural identity)*
> ***Brainstormed Connections***
> *It's interesting that often times one's gender can influence what society thinks is "appropriate" for them to do in their professional life. How does one's gender and society's construction of gender affect what someone does with their professional life?*
> ***Integrated Question***
> *In what ways has society's view of you as a woman or man affected the path that you chose to go down in your professional life? (**Constructs**: expectation, society, gender)*

These multiple ways of questioning her micro stories have allowed A. to make her implicit questions visible. Instead of staying stuck in her eddy of sentences declaring that she does not know how to answer important questions in her life, she is reframing these questions, directing them to others, and later redirecting them to herself.

"What's Your Story?": Gaining Self-authorship

We offer a case study of student C. from the first iteration of our course as an example of a more processed work integrating personal and professional through the lens of gender and relatedness. In her final work, C. has come to overtly examine her "personal" self-expression of gender within the context of her engineering classroom, and to project the desire for more connectivity between the two:

> *When I walk into the room, the first thing I notice is how much effort I have to extend to pick out the females from the males. It is a sea of T-shirt, khakis, faded denim, and hiking boots. In my black wedge heels, short lace dress, and red lip, I recognize that I stand out like a sore thumb. As I walk, I turn heads. I don't say this, because I think I am anything particularly worth looking at. I say this, because as a fashionable woman who attends a school of full-time engineers, I am aware that I wear my femininity more loudly than my peers. This gets me more than the few odd looks.*
>
> ...
>
> *You see, to me, clothing is more than a basic need; it is a form of self-expression. I take pride in my feminine shape and my ability to communicate my personality in such a visual and tactile format. I want my external presentation to fully express how I feel. To me it seems like the most natural view of fashion, but for some reason, at a school of engineers, this makes me odd.*
>
> ...
>
> *I want to be an engineer and take pride in my femininity. I want to display it proudly, wear it draped on my form with the same comfort I wear the title of engineer.*

C.'s work evolved over the course of the semester from compartmentalization (my interest in fashion has no place within the stereotypically unfeminine engineering community) to a manifesto for integration: "I want to be an engineer and take pride in my femininity. I want to display it proudly, wear it draped on my form with the same comfort I wear the title of engineer." With our literary criticism and critical discourse analysis lens, we note the merging of engineering with femininity—"I want to be an engineer *and* take pride in my femininity," and we can observe how the language of feminized attention to fashion bleeds into the language of engineering—C. wants to "wear" the title of engineer. In this sense C.'s narrative has powerfully begun to enact what it hopes for: she is finding language and connections that bridge her gendered self-expression in the realm of fashion with her self-expression in the world of engineering. (And, in fact, C.'s iterative composition of this piece inspired her to examine how she could bring her love for constructing clothing into her engineering projects, i.e., the process of clothes design has begun blending with an engineering design process and enterprise).

However, by no means does this narrative make tidy work of the complex questions that arise from critical reflection in the context of relatedness and gender. C. expresses numerous challenges with relatedness, most specifically in terms of her disconnection from engineering peers: she characterizes herself as someone who wears her femininity "loudly," rendering her a "sore thumb" who "sticks out" with her "red lip" in contradistinction to an undifferentiated "sea" of less colorful ("khaki," "denim") cohorts. While C. aims at the end of the piece to feel proud of the "display" she creates, one could also wonder about the toll that feeling so different creates: whereas she perceives her classmates as a unified and difficult to distinguish, C.'s descriptors posit her as an uncomfortable ("sore") exception. But like her red lip, this uneasiness has become visible and perhaps even beautiful: to her instructors, to her peers, and to herself.

And this is the project of critical reflection. Like C., we aim for a vision statement where students, for having explored them, will wear their unique discoveries about themselves both more comfortably and more proudly.

"What's Your Story?": What About Us?

As evident in our exploration of the stories A. and C. develop and share, the act of critical reflection is neither tidy nor finite: the project of unearthing, interrogating, and sharing uneasy stories is a rewarding but messy one that requires iteration across a lifetime. We ask that our students take up this ongoing project, encouraging them to step into the vulnerable spotlight of critical self-reflection within and beyond their college years and we ask them to do so "here and now" within their engineering work, engineering curriculum, and in the process of engineering their lives.

But what about the rest of us? In our learning communities, how can we ask students to excavate and communicate their own stories without walking the walk of unpacking and sharing "uneasy" stories of our own? How many of us, like A., have been sitting on the sidelines of our professional communities and failing to say or perhaps even know what core values and stories implicitly drive our teaching, our connection with students, our connection to colleagues, our publications, and our intellectual pursuits and ambitions? The unspoken politics of "professionalism" perhaps encourage us to remain silent even as our students engage in profoundly important conversations about what they care about. And the immense time commitment of academic teaching and research perhaps dissuades us from dusting the cobwebs off the teaching and learning statements written by our younger selves, or wrestling more analytically with our requisite stories and insights in progress reports and promotion processes.

We believe our students will take more and better intellectual and emotional risks if we model that behavior in our classroom communities, our academic institutions, and our outward-facing narratives—meaning that we too must also step away from the sidelines and into the action. We too must perform integration of the personal pattern of knowing (Carper, 1978) into our professional practice and art of ethical coaching, mentoring, advising, supporting, developing, and simply "being" with the students. We too must practice openness about our own professional and personal unstable equilibria (or absence of equilibria) and allow for an open conversation about "movement toward growth and fulfillment … and] the recognition that each person is not a fixed entity, but constantly engaged in the process of becoming" (Carper, 1978). We too must acknowledge vulnerabilities involved in the hard work of critical reflection and the unique ways in which authentic practice of critical reflection allows for building professional and personal integrated selfhood with personal, ethical, empirical, and aesthetic fundamental patterns of knowing (Carper, 1978). We too must know ourselves to relate to our students—this is the only way to grow our own creative ways of knowing to allow for growing our students' creative ways of knowing.

To that end, we want to share a bit of our respective stories with you: underscoring what motivates us to promote critical reflection in academia, and how such reflection transformed our two specific stories of gendered disconnection in professional spheres into an integrated vision for change.

"What's Your Story?": Breaking the Faculty Fourth Wall

Gillian Epstein, Ph.D. in English Literature from UC Berkeley

In the land of the free, I grew up in a domestic dictatorship governed by fear in the form of a rigidly authoritarian woman who snapped into seismic emotional rages when a kid stepped outside the party line. As in most emotionally abusive households, the line was arbitrary, and probably impossible to follow even if one had a compliant nature. Which I did not. I spent the bulk of my younger years exiled to my room, losing myself in novels beyond my years and writing notes to my mother on scraps of pink stationery which I would then slip into the void under her closed bedroom door, delivering analytical scrawls in bold felt-tip marker challenging her unwavering story that I was born bad.

So associatively, coming of age in the "soft" world of an elite English Literature doctoral program was safe territory—a place where I could continue to cultivate my literary eye and analytical voice, but within a fixed intellectual framework that would not disturb any dark spirits lurking in my childhood bedroom. In this hard, soft world it was most certainly okay to study and write about topics that associatively might matter to one's personal subjectivity, such as feminist, postcolonial, or queer theory. But it was most certainly not okay to make those connections explicit by overtly processing stories of our own: an implicit intellectualized don't ask, don't tell policy. I happily spoke this "objective" literary theory and criticism language with fluidity, excelled in my courses and sped through the qualifying process for writing a dissertation. That is, until I ran straight into an unanticipated brick wall named the *Portrait of a Lady*: a canonical nineteenth-century realist novel by Henry (1963) that tells the story of Isabel Archer—a bright, independent young woman articulately intent on blazing a trail beyond domesticity but who inexplicably ends up in an emotionally abusive marriage with a hyperbolically controlling man, ultimately losing her agency and voice. Unfortunate, but not unexpected: there are piles of novelized, nineteenth-century cautionary tales retracing the path of a good girl gone wrong and this was not my first time at the reading rodeo.

So I was taken by surprise that when I finished the book I felt as undone as its heroine. Instead of writing up notes, I cried. When I attended class the next day, I was suddenly incapable of making my usual prizefighting analytical comments. "Why," I kept asking, "would James *do* that to Isabel Archer? What kind of sadistic freak builds up the most appealing, smart, and interesting nineteenth-century American heroine that most of us have ever run across just to slowly bleed the life out of her? This book is *heartbreaking*!" The female professor and the rest of the class looked away uncomfortably, as if I had a strange flu. Like them, I figured my immunity to the perils of plot would kick in sooner than later. But it didn't. I buried myself in James's famous preface to the book and got an inkling of why: the paradoxical conceit of this novel is that its heroine has attempted feminized self-authorship and failed, and that the male narrator and author had no hand in literally plotting against her.[2] Subtextually, the restrictive framework of gender had eerily mapped itself along the lines of my childhood outrage about iniquitous narration: a controlling and condescending parent stifling a child's self-realization and saying it is due to the child's flawed nature; an all powerful and patronizing author stifling a heroine's infantilized attempt at independence and ascribing it to her feminized fallibility. I needed to undo the narration and ideology of Isabel's undoing so that as a woman I had a chance—as James might condescendingly put it—to "matter."[3]

I made the mistake of telling my advisor the emotional motivation behind my work; overnight, she suddenly went from a full year of supporting my project to asking me to repeat the onerous approval process. I switched advisors, and this time was careful to keep my mouth shut about my agenda and to cushion the sharp challenge of my analysis in bubble wrapped layers of theoretical jargon. But despite my outward compliance with codes, I knew I had failed the test of keeping art and life separate. I promised myself that if I could just finish this project and secure my doctorate, that I would never again write a piece of academic work where I so absurdly hid my voice and motivations. Which I figured meant never writing another academic piece in my discipline again. Which meant that even though I had graduated with honors from the highest-ranked English doctoral program in the country, I chose not go on the academic job market for an English faculty job since they would all require research publications. Instead, I went rogue and joined an engineering college as the only Humanities Ph.D. on staff, and was tasked with "integrating" the humanities across the engineering curriculum, which posed an interesting conundrum. Here were a majority of students (plus many faculty) who saw reading, writing, and analyzing stories as a frivolous waste of time: unless I could make a compelling, explicit case about why any of them should personally care about how stories shape us and how they are told, I was history. Ironically, joining the hard sciences as a self-banished practitioner from the sterile, "soft" Humanities necessitated that I create conversations about personal connections (and disconnections) with narrative. A stranger in a strange land, I am uneasy at home: I can use my voice, challenge dominant narrations, and help students do the same. And that is why, after so many years of professional silence, I am free to write this academic piece for you, charming reader.

[2] "…the germ of my idea…consisted not at all in any conceit of a 'plot,' nefarious name….but altogether in the sense of a single character, the character and aspect of a particular engaging young woman…" (James, 4–5).

[3] "Millions of presumptuous girls, intelligent or not intelligent, daily affront their destiny, and what is it open to their destiny to be…that we should make an ado about it?....how absolutely, how inordinately, the Isabel Archers, and even much smaller female fry, insist on mattering…." (James, 9, 11)

Yevgeniya V. Zastavker, Ph.D. in Biophysics from MIT

As I stood on the platform of the main Kiev Rail Station taking in the last few breaths of the radioactive air, I kept thinking that this was the last time I was "othered." It was July 1989. I was leaving behind a place where the main rubric I was measured by was "'Paragraph No. 5:' Nationality: Jewish," a rubric that forced me to graduate from secondary school 2 years earlier, a rubric that banned me from applying to the university I wanted to apply to, a rubric that forced me to work ten times harder than anybody else just to be recognized. Paragraph No. 5 inscribed my future, chiseled my next steps, and silenced my voice, but if I stayed quiet, really-really quiet, and my passport wasn't necessary, I could just get by, a shadow among those who proudly carried "Nationality: Ukrainian." And as I stood there drinking in the last few sips of that radioactivity, I was imagining how it would be once I left this country, how it would feel to learn about and gain my voice, how it would feel to be free, to be empowered, to be recognized, to be a part of an inner circle in a society where meritocracy is the way of living and equity is the way of being.

Ten years later, almost to date, I was sitting in a large brightly lit graduate student office at MIT. I was a physics student at MIT! MIT! Who could have thought?! There were just two of us in that office, myself and my co-advisor. With the office door closed shut, he pulled my chair close to his so that our knees almost touched and I felt his breath on my face. In a half whispered voice he told me, "Zhenya," a short version of my name, "Your project is trash; your work is trash; you should collect it and throw it out." He pointed to the blue-gray trashcan next to the black office door. "Go home and wash dishes," he added. Silence. As I was sitting there, my knees close to his, I was breathing in that hot radioactive air all over again. I was "othered" again. You see, in the Soviet Union I could keep my Paragraph No. 5 hush-hush, at least among people who did not know me; but there was nowhere to hide being a woman at MIT in the physics department. I was obvious; I was in everybody's face; I did not belong. Meritocracy-schmeritocracy. I still needed to work ten times harder just to make it, this time because of Paragraph No. 2: Female. My future was inscribed again. Or was it?

Five years later, I was sitting in a large brightly lit faculty office, my office. I became a faculty member! No, I didn't just "become" faculty. I earned this place! I earned it with hard work, I earned it despite others, I earned it because I wanted to help clear the road for other young women who deserve to follow their passion to do math, science, or engineering in a meritocratic society. I imagined myself to be Joan of Arc who proudly carries a flag of womanhood in a masculinized battle culture of academia and I imagined leading many young women forward to the siege of their imagined futures in science and engineering. I gained my voice; I heard it loudly and clearly, and I remained "othered" this time around by my own choice. And I am not necessarily talking about being a *science* faculty member at an *engineering* college although it plays a role. And I am not necessarily talking about being a *woman* in a hard domain of engineering although that necessarily plays a role. I am talking about going rogue and focusing on *humanities*, specifically, the discipline of education as my predominant intellectual activity, research, practice, and life. Ironically, integrating the "soft" Humanities into "hard core" engineering education made me free to talk about and practice the art of teaching and learning, the craft of being a woman in an engineered world of masculinity, and the aesthetics of nurturing our students' selfhoods. By doing so, I have also narrated myself to become a healer rather than a warrior, a much more quiet yet, in my mind, critically important place for a faculty member whose main responsibility is to work with students in co-creation of a culture of learning, "where wicked problems are tackled together in a spirit of mutual learning and understanding" (Rolfe, 2014). As Gillian, I am a stranger in a strange land, but I am finally at home.

References

Adams, R. S., Turns, J., & Atman, C. J. (2003). Educating effective engineering designers: The role of reflective practice. *Design Studies, 24*, 275–294.

Ahern, A., O'Connor, T., McRuairc, G., McNamara, M., & O'Donnell, D. (2012). Critical thinking in the university curriculum—The impact on engineering education. *European Journal of Engineering Education, 37*, 125–132.

Alsup, J. (2013). *Teacher identity discourses: Negotiating personal and professional spaces.* New York, NY: Routledge.

Andrews, T., & Patil, R. (2007). Information literacy for first-year students: An embedded curriculum approach. *European Journal of Engineering Education, 32*, 253–259.

Blatt, S. J. (1974). Levels of object representation in anaclitic and introjective depression. *Psychoanalytic Study of the Child, 29*, 107–157.

Blatt, S. J. (1995). Representational structures in psychopathology. In D. Cicchetti & S. Toth (Eds.), *Rochester Symposium on Developmental Psychopathology* (Emotion, cognition, and representation, Vol. 6, pp. 1–33). Rochester, NY: University of Rochester Press.

Blatt, S. J. (1998). Contributions of psychoanalysis to the understanding and treatment of depression. *Journal of the American Psychoanalytic Association, 46*, 723–752.

Blatt, S. J., & Blass, R. B. (1996). Relatedness and self definition: A dialectic model of personality development. In G. G. Noam & K. W. Fischer (Eds.), *Development and vulnerabilities in close relationships* (pp. 309–338). Hillsdale, NJ: Erlbaum.

Blatt, S. J., & Shichman, S. (1983). Two primary configurations of psychopathology. *Psychoanalysis and Contemporary Thought, 6*, 187–254.

Blatt, S. J., & Zuroff, D. C. (1992). Interpersonal relatedness and self-definition: Two prototypes for depression. *Clinical Psychology Review, 12*, 527–562.

Bruner, J. (2002). *Making stories: Law, literature, life*. New York: Farrar, Straus and Giroux.

Bruno, B. A., & Anderson, A. (2005). Using objective driven heat transfer lab experiences to simultaneously teach critical thinking skills and technical content. In *Proceedings of American Society of Mechanical Engineers (ASME) International Mechanical Engineering Congress and Exposition (IMECE).*

Carper, B. A. (1978). Fundamental patterns of knowing in nursing. *Advances in Nursing Science, 1*(1), 13–24.

Catalano, G. D. (1993). Developing an environmentally friendly engineering ethic: A course for undergraduate engineering students. *Journal of Engineering Education, 82*, 27–33.

Chang, P.-F., & Wang, D.-C. (2011). Cultivating engineering ethics and critical thinking: A systematic and cross-cultural education approach using problem-based learning. *European Journal of Engineering Education, 36*, 377–390.

Choulier, D., Picard, F., & Weite, P.-A. (2007). Reflective practice in a pluri-disciplinary innovative design course. *European Journal of Engineering Education, 32*, 115–124.

Claris, L., & Riley, D. (2012). Situation critical: Critical theory and critical thinking in engineering education. *Engineering Studies, 4*, 101–120.

Claris, L., & Riley, D. (2013). Situation critical: Critical theory and critical thinking in engineering education. *IEEE Women in Engineering Magazine*, 32–36.

Cloete, A. (2001). Solving problems or problem solving: What are we teaching our students? In *Proceedings of 2001 American Society of Engineering Education Annual Conference.*

Consortium to Promote Reflection in Engineering Education (CPREE). University of Washington. Retrieved from http://cpree.uw.edu/

Deci, E., & Ryan, R. (Eds.). (2002). *Handbook of self-determination research.* Rochester, NY: University of Rochester Press.

Dewey, J. (1910). *How we think.* Boston: D.C. Heath and Company.

Dewey, J. (1933). *How we think: A restatement of the relation of reflective thinking to the educative process.* Boston: D.C. Heath and Company.

Doel, S. (2009). Fostering student reflection during engineering internships. *Asia-Pacific Journal of Cooperative Education, 10*(3), 163–177.

Donawa, A., Martin, C., & White, C. (2007). Re-engineering engineering: Teaching students how to think critically. In *Proceedings of 2007 American Society of Engineering Education Annual Conference.*

Facione, P. A. (1990). *Critical thinking: A statement of expert consensus for purposes of educational assessment and instruction. Research Findings and Recommendations.* Retrieved from http://www.eric.ed.gov/ERICWebPortal/detail?accno=ED315423/

Facione, P. A. (1998). Critical thinking: What it is and why it counts. Millbrae, CA: California Academic Press. Retrieved February 26.

Fairclough, N. (1995). *Critical discourse analysis: The critical study of language.* London: Longman.

Franklin W. Olin College of Engineering at www.olin.edu

Fook, J. (2013). Critical reflection in context contemporary perspectives and issues. In J. Fook & F. Gardner (Eds.), *Critical reflection in context of applications in health and social care* (pp. 1–11). London: Routledge.

Gardner, F. (2011). *Critical spirituality: A holistic approach to contemporary practice.* Farnham: Ashgate.

Garnder, F. (2014). *Being critically reflective: Engaging in holistic practice.* New York, NY: Palgrave Macmillan.

Henry, J. (1963). *The portrait of a lady* (1908th ed.). In L. Edel (Ed.) Boston: Houghton Mifflin.

Hicks, N., Bumbaco, A. E., & Douglas, E. P. (2014). Critical thinking, reflective practice, and adaptive expertise in engineering. In *Proceedings of 2014 American Society of Engineering Education Annual Conference.*

Hickson, H. (2011). Critical reflection: Reflecting on learning to be reflective. *Reflective Practice, 12*(6), 829–839.

Hunter, A.-B., Laursen, S. L., & Seymour, E. (2007). Becoming a scientist: The role of undergraduate research in students' cognitive, personal and professional development. *Science Edition, 91*, 36–74.

Jacob, S. A. (2012). Writing interview protocols and conducting interviews: Tips for students new to the field of qualitative research. *The Qualitative Report, 17*(6), 1–10.

Johns, C. (2005). Reflection on the relationship between technology and caring. *Nursing in Critical Care, 10*(3), 150–155.

Khisty, C. J., & Khisty, L. L. (1992). Reflection in problem solving and design. *Journal of Professional Issues in Engineering Education and Practice, 118*, 234–239.

Lin, X., Schwartz, D. L., & Bransford, J. (2007). Intercultural adaptive expertise: Explicit and implicit lessons from Dr. Hatano. *Human Development, 50*, 65–72.

Lombardo, S. J. (2004). Using small blocks of time for active learning and critical thinking. *Chemical Engineering Education, 38*(2), 150–153.

Keegan, R. (2002). Step-back consulting (background from the Harvard Macy Institute). Retrieved from http://transformationalschoolleadership.com/wp-content/uploads/Step-Back-Consulting-Process.pdf

McKenna, A. F. (2007). An investigation of adaptive expertise and transfer of design process knowledge. *Journal of Mechanical Design, 129*, 730–734.

McKenna, A. F., Yalvac, B., & Light, G. J. (2009). The role of collaborative reflection on shaping engineering faculty teaching approaches. *Journal of Engineering Education, 98*, 17–26.

Miller, R. L., & Olds, B. M. (1994). A model curriculum for a capstone course in multidisciplinary engineering design. *Journal of Engineering Education, 83*(4), 311–316.

Mina, M., Omidvar, I., & Knott, K. (2003). Learning to think critically to solve engineering problems: revisiting John Dewey's ideas for evaluating the engineering education. In *Proceedings of 2003 American Society of Engineering Education Annual Conference.*

Mina, M., Cowan, J., & Heywood, J. (2015). Case for reflection in engineering education – and an alternative. *In Frontiers in Education Conference Proceedings.*

National Academy of Engineering. (2009). *Grand Challenge Scholars Program.* Retrieved from http://engineeringchallenges.org/14365/GrandChallengeScholarsProgram.aspx

National Academy of Engineering. (2004). *The engineer of 2020: Visions of engineering in the new century.* Washington, DC: National Academies Press.

National Academy of Engineering. (2005). *Educating the engineer of 2020: Adapting engineering education to the new century.* Washington, DC: National Academies Press.

Papadopoulos, C., Rahman, A., & Bostwick, J. (2006). Assessing critical thinking in mechanics in engineering education. In *Proceedings of 2006 American Society of Engineering Education Annual Conference.*

Paul, R. (2004). *The state of critical thinking today.* Retrieved from http://www.criticalthinking.org/pages/the-state-ofcritical-thinking today/523

Paul, R., Elder, L., & Foundation for Critical Thinking. (2003). *The miniature guide to critical thinking: Concepts & tools.* Dillon Beach, Calif: Foundation for Critical Thinking.

Phillips, C., & Green, S. (2011). Faculty as critical thinkers. *Inquiry: Critical Thinking Across the Disciplines, 26*(2), 44–50.

Polanco, R., Calderon, P., & Delgado, F. (2001). Effects of a problem-based learning program on engineering students. In *Proceedings of 2001 American Educational Research Association.*

Rayne, K., Martin, T., Brophy, S., Kemp, N. J., Hart, J. D., & Diller, K. R. (2006). The development of adaptive expertise in biomedical engineering ethics. *Journal of Engineering Education, 95*, 165–173.

Riley, D. (2003). Employing liberative pedagogies in engineering education. *Journal of Women and Minorities in Science and Engineering, 9*(2), 137–158.

Riley, D. (2008). *Engineering and social justice.* San Rafael, CA: Morgan and Claypool.

Riley, D. (2015). Facepalms and cringes: Liberal education misapprehended. *Engineering Studies, 7*(2–3), 138–141.

Riley, D., & Claris, L. (2006). Power/knowledge: Using Foucault to promote critical understandings of content and pedagogy in engineering thermodynamics. In *Proceedings of 2006 American Society of Engineering Education Annual Conference.*

Rittel, H., & Webber, M. (1973). Dilemmas in a general theory of planning. *Policy Sciences, 4*, 155–169.

Rolfe, G. (2002). Reflective practice: Where now? *Nurse Education in Practice, 2*, 21–29.

Rolfe, G. (2014). Rethinking reflective education: What would Dewey have done? *Nurse Education Today, 34*, 1179–1183.

Romkey, L. (2009). The development and assessment of critical thinking for the global engineer. In *Proceedings of 2009 American Society of Engineering Education Annual Conference.*

Ryan, R. M., & Deci, E. L. (2000). Self-determination theory and the facilitation of intrinsic motivation, social development, and well-being. *American Psychologist, 55*, 68–78.

Schön, D. A. (1983). *The reflective practitioner: How professionals think in action.* New York: Basic Books.

Schön, D. A. (1987). *Educating the reflective practitioner: Toward a new design for teaching and learning in the professions.* San Francisco: Jossey-Bass.

Sepp, L. A., Orand, M., Turns, J. A., Thomas, L. D., Sattler, B., & Atman, C. J. (2015). On an upward trend: Reflection in engineering education. In *Proceedings of 2015 American Society of Engineering Education Annual Conference.*

Shahar, G., Henrich, C. C., Blatt, S. J., Ryan, R., & Little, T. D. (2003). Interpersonal relatedness, self-definition, and their motivational orientation during adolescence: A theoretical and empirical integration. *Developmental Psychology, 39*(3), 470–483.

Swalie, B. H., & Kreppel, M. C. (2001). Building critical thinking, teamwork, and communication skills through professional ethics in engineering and chemical technology. In *Proceedings of 2001 American Society of Engineering Education Annual Conference.*
Takacs, D. (2003). How does your positionality bias your epistemology? *The NEA Higher Education Journal,* 27–38.
Turns, J. (1997). Learning essays and the reflective learner: Supporting assessment in engineering design education. *In Frontiers in Education Conference Proceedings.*
Turns, J. A., Sattler, B., Yasuhara, K., Borgford-Parnell, J., & Atman, C. J. (2014). Integrating reflection into engineering education. In *Proceedings of 2014 American Society of Engineering Education Annual Conference.*
U.S. News and World Report. (2015). Education. Retrieved from http://colleges.usnews.rankingsandreviews.com/best-colleges/olin-college-39463
Van den Broeck, A., Vansteenkiste, M., De Witte, H., Soenens, B., & Lens, W. (2010). Capturing autonomy, competence, and relatedness at work: Construction and initial validation of the work-related basic need satisfaction scale. *Journal of Occupational and Organizational Psychology, 83,* 981–1002.
Vos, H., & de Graaff, E. (2004). Developing metacognition: A basis for active learning. *European Journal of Engineering Education, 29,* 543–548.
Walther, J., Radcliffe, D., & Mann, L. (2007). Analysis of the use of an accidental competency discourse as a reflective tool for professional placement students. *In Frontiers in Education Conference Proceedings.*
Walther, J., Kellam, N., Radcliffe, D., & Boonchai C. (2009). Integrating students' learning experiences through deliberate reflective practice. *In Frontiers in Education Conference Proceedings.*
Walker, J. M., Cordray, D. S., King, P. H., & Brophy, S. P. (2006). Design scenarios as an assessment of adaptive expertise. *Situations, 1,* 2.
Walther, J., Sochaka, N. W., & Kellam, N. N. (2011). Emotional indicators as a way to initiate student reflection in engineering programs. In *Proceedings of 2011 American Society of Engineering Education Annual Conference.*
Wilson, D., Jones, D., Kim, M. J., Allendoerfer, C., Bates, R., & Crawford, J. (2014). The link between co-curricular activities and academic engagement in engineering education. *Journal of Engineering Education, 103,* 625–651.
Wolverton, R. H., & Wolverton, J. B. (2003). Implementation of ethics education throughout an engineering college. In *Proceedings of 2003 American Society of Engineering Education Annual Conference.*
Zastavker, Y. V., & Epstein, G. (2015). Critical reflection across Olin's learning continuum. *Internal Report for Curriculum Innovation Grant Program.*

Example Mission Statements

University of Maryland, The Department of Mechanical Engineering Mission Statement at http://www.enme.umd.edu/about/mission
Lamar University, College of Engineering at http://engineering.lamar.edu/chemical/mission-statement.html
The University of Southern Mississippi, School of Construction at http://www.usm.edu/construction/mission-statement-0
University of Pennsylvania, School of Engineering and Applied Sciences at http://www.seas.upenn.edu/education/
Bucknell University, The Department of Mechanical Engineering at http://www.bucknell.edu/college-of-engineering/enrollments-graduation-rates-and-educational-outcomes/abet-mechanical-engineering.html

Ethical Dilemmas in the Engineering Writing Classroom

Kevin C. Moore

Introduction

When it comes to promoting creativity in university engineering education, ethics may not be the field that most readily comes to mind. For engineers, ethics usually exists as an arena of potential legal, business, social, and political complications, where specialized technical advancement proceeds with, hopefully, as little friction as possible. Ask an engineer what it is that engineers *do*, and they will all too frequently tell you that engineering has to do with the "how" rather than the "why." Engineers may solve problems, but they do so mainly, in their perception, as agents of corporate, government, and academic entities. Stereotypically speaking, ethics, including questions of consequence and responsibility, may seem a regulating factor for engineering ingenuity, perhaps even a limiting factor. To deploy an apt metaphor, ethical discussions and assignments are often seen as "the brakes" for engineering creativity. Ethics appears as an afterthought in engineering rather than a primary concern; however paradoxical this may be for a field with the ostensible goals of streamlining and improving the world.

If ethics is one set of "brakes" for engineers, university writing requirements may be another, at least in the perception of engineering students. At institutions with WID/WAC programs offering engineering writing courses, however, such as the University of California, Santa Barbara (UCSB), where I teach,[1] dedicated engineer-

[1] I am a Lecturer in the UCSB Writing Program, which, as of academic year 2016–2017, is an independent campus unit of 47 full-time faculty. The Writing Program offers undergraduate general education courses at the lower and upper division levels, in discipline-specific as well as interdisciplinary and professional writing contexts; a professional writing minor; graduate courses focusing on writing and writing pedagogy; and a Ph.D. emphasis in Writing Studies. The Engineering Writing Sequence (or "E-Sequence") is a dedicated lower-division series of three courses, which exclusively serve students in UCSB's College of Engineering in their first 2 years of study.

K.C. Moore (✉)
University of California, Santa Barbara, Santa Barbara, CA 93106, USA
e-mail: moore@writing.ucsb.edu

D. Bairaktarova, M. Eodice (eds.), *Creative Ways of Knowing in Engineering*,
DOI 10.1007/978-3-319-49352-7_8

ing writing courses also become spaces where the role of ethics in engineering education can be reconsidered. This is in part because of smaller class sizes—at UCSB, all undergraduate writing courses are strictly capped at 25—and the interactive nature of contemporary writing pedagogy. Even in courses that foreground technical or professional genres (technical reports, proposals, correspondence, web design, job materials, etc.), the processes of collaboration—peer collaborations between students, as well as individual and group collaborations between students and their instructors—make the writing classroom a space where ethical questions can be voiced serendipitously, frankly, and speculatively. Writing courses moreover often facilitate "low-stakes" writing activities, including drafting and revision, as part of their effort to help students build not just acumen, but confidence and perspective. This disrupts the product obsession that largely defines our culture of technological progress, offering opportunities for students to rethink their analytic and creative processes. The writing classroom produces a relative "safe space" within the undergraduate engineering experience, where ethical questions, including questions that might make some engineers uncomfortable, can be raised with greater frequency and candor than in large lecture courses, or laboratory settings.

The significant sociopolitical, technological, and environmental challenges humanity faces at the beginning of the twenty-first century, and perhaps especially the major ethical questions surrounding sustainability, should have a meaningful role in the education of engineers. But only when presented as an essential—rather than supplemental—component of engineering epistemology and practice can ethics take on such a role. The engineering writing classroom itself may seem, to some, as a further supplementary space within the general engineering curriculum, but its relative intimacy, as well as its focus on collaboration and process, allows instructors to prototype what ethics education could look like in general engineering contexts. In this study, I examine the role of ethics in the engineering writing classroom theoretically as well as through real-world examples, presenting the writing classroom as a proving ground for rethinking the role of ethics in engineering education. I begin by sketching the background of ethics in engineering education, responding to recent calls to action along these lines. I then turn to Steven B. Katz's (1992) influential essay "The Ethic of Expediency: Classical Rhetoric, Technology, and the Holocaust," which illustrates why ethics need to take on a central and in fact an urgent role in engineering, exploring how a rhetorical model for approaching ethics might improve the quality of engineering ethics education and also constitute a creative engagement. This, I argue, should be the primary goal of engineering ethics education: to ask students to consider ethics not as supplementary, but as an essential part of engineering, and in fact as a medium of creativity. I then explore some of the pitfalls of an especially fragmented approach to ethics in a widely used engineering writing textbook, David F. Beer and David A. McMurrey's (2014) *A guide to writing as an engineer* (4th ed.) to illustrate how ethics too often takes on the role of a supplemental priority for engineers, even in the midst of what presents itself as a dedicated engagement of the ethical dimension. I also investigate the larger theoretical issue of why this is so, citing three reasons for the marginal status of ethics in engineering education—the complexity of predicting how a given technology will evolve and adapt, the major differences in ethical considerations between engineering subfields, and the individual engineer's limited agency—and furthermore respond to imperatives by Erin A. Cech (2014), Olivia Walling (2015), and a new study by Diego Gambetta and Steffen Hertog (2016) suggesting that limited ethical training may be a serious professional and personal liability for engineers. Finally, I present primary research conducted in an engineering writing course at UCSB involving creative ethics assignments, which require students to defend potentially problematic technologies as a way of developing an awareness of how ethical considerations can generate legitimate and innovative ideas for new engineering solutions. When ethics become a domain of creativity, I argue, engineering students are hardly constrained—in fact they can thrive.

Background

Since the last quarter of the twentieth century, the Accreditation Board for Engineering and Technology (ABET) and institutions of higher learning alike have recognized an urgent necessity to provide engineers with ethical training, generally in the form of a dedicated undergraduate engineering ethics course (Harris, Davis, Pritchard, & Rabins, 1996; Harris, Pritchard, & Rabins, 1996; Herkert, 2000). This has been in part a response to historical instances where engineers have been involved in the development of technologies that have threatened human civilization on a large scale, such as the development of nuclear energy and weapons, as well as increased public awareness about the state of the environment and sustainability practices (Herkert, 2000). While not as directly relevant for engineers as for medical and social sciences researchers, increased human subjects protections on university campuses, and the rising authority of the Institutional Review Board, also constitute backgrounds. In academic contexts, increased attention to ethics in engineering departments is part of a generally increased attention to ethical research. Highly politicized, public engineering-related concerns—including genetic engineering, cybernetics, and artificial intelligence research—also provide a context for the increasingly urgent attention placed on ethics by engineering educators. Recently, there has moreover been alarm among sociologists researching engineering education over the extent to which current practices may create a "culture of disengagement" (Cech, 2014), or even incubate a dangerous psychology prone to ideological activation, specifically jihadist terrorism (Gambetta & Hertog, 2016). The mechanized genocide of the Holocaust looms as a larger context here as well, and what Katz (1992) calls "the ethic of expediency," a dominant rhetorical paradigm with roots extending to Ancient Greece and the very origins of Western philosophy.

I attend in depth to these specific, nuanced debates below. At the outset of this account, let it suffice to assume that, despite a growing body of research on effective engineering ethics pedagogy, and considerable debate regarding best approaches to effective practices, ethics remains a supplemental, fragmented, and often inherently contradictory component of engineering education. As Olivia Walling (2015) points out, one limitation of current practices of teaching engineering ethics is a focus on cognitive learning, mainly through case studies. Most textbook authors addressing engineering ethics pedagogy have tried to limit the field's domain to "professional ethics" (Beer and McMurrey, 2014; Harris, Davis, et al., 1996; Harris, Pritchard, et al., 1996; Mitcham & Duvall, 2000), claiming that students arrive at the university "more or less morally mature" (Harris, Davis, et al., p. 93). Yet researchers have found it difficult to prove conclusively that ethics education actually makes a meaningful impact in ethical decision-making (Hashemian & Loui, 2010), and have largely limited their assessment of ethical learning to whether students achieve business ethics skills such as "professional responsibility" (p. 215). Innovative new methods, including Walling's (2015) attempt to ask students to conduct and reflect upon a series of ethical experiments in the "stressful and competitive environment that is part of engineering student life" (p. 1644), certainly go further in this regard, providing a more balanced and effective model. But a lingering preoccupation with expedient ethical training—where ethics education is taught as something supplemental to the real work of engineering—remains the status quo, even in otherwise innovative and useful educational materials such as Beer and McMurrey's (2014) textbook *A Guide to Writing as an Engineer* (4th ed.), where ethics is treated late in the volume as a peripheral topic.

The general institutional approach to ethics as supplementary is a problem, because, as Erin A. Cech (2014) succinctly puts it, "If engineers cannot adequately reflect upon the social impact of their work, there are few individuals in the lay public with the specialized competencies to do so" (p. 44). Proceeding from a writing studies perspective, but with a broader interest in the role of ethics in engineering education generally, this study follows recent imperatives by Cech and Walling to examine how instructors may more effectively raise ethical questions not only to promote critical thinking, but to explore ethical problems as legitimate opportunities for rhetorical and technical ingenuity. It proposes, moreover, that the perceived supplemental status of ethics training for engineers constitutes an impediment not only to sound practices of business and innovation, but to creativity itself. When students are presented with troublesome ethical dilemmas not as problems to solve or impediments to the real engineering "work" of technological innovation, but as a *medium* of technological invention—not just as complications and background noise, but part of the epistemology of engineering as a field—they experience opportunities to rethink the larger stakes of engineering endeavors. This requires going beyond merely presenting students with case studies, which, as Walling (2015) points out, depict ethics as primarily an exercise in analytic reasoning. According to Walling, the practices of using case studies in ethical education emerge from legal instruction rather than engineering itself, where it constitutes an "import," and an insufficient one. There is an important difference here: when law "students analyze cases, they are practicing law because it is the work of lawyers to analyze texts and to develop understandings of them in order to create persuasive texts. Texts are not the object of study or work product in engineering" (p. 1648). In contrast, by confronting rather than ignoring the limited ethical agency engineers possess while working in collaborative settings, whether in the public or private sector, it becomes possible to foster a realistic perspective on the field of engineering, its possibilities, and the goals of engineers at the outset of their careers.

Ethical engagements in engineering education typically take on a form where students approach and write about ethics as engineers, but with the final assumption that ethics is a field apart from engineering. What would it mean to ask engineering students to consider ethics an essential part of all of their work, to think of engineering creativity as an implicitly ethical field: to not only engineer more ethically, but to engineer solutions to ethical problems? The goal of engineering ethics education should be to cultivate an awareness of engineering work as inherently ethical in nature, to recognize the danger of marginalizing ethical priorities.

"The Ethic of Expediency" and Deliberative Rhetoric as a Pathway to Ethical Creativity

It may seem extreme to present the Holocaust as an essential background to consider when trying to understand the increased attention to ethics in engineering since the second half of the twentieth century. But if, as the German philosopher and sociologist Theodor Adorno (1949/1997) famously wrote, "To write poetry after Auschwitz is barbaric" (p. 34), then the specter of barbarism hangs even higher over post-Holocaust engineering and technical communication. What Katz (1992) calls the predominant "ethic of expediency" in Western culture, especially as it manifests itself in technical communication, may nevertheless provide a medium for engineers to engage, and one with the potential to provide engineers a means of not simply recognizing ethics as an important supplementary field but a domain of essential importance to creativity.

According to Katz (1992), one way to understand "that question murmured so many times before—how could the holocaust [sic] have happened?" is to recognize that "the imperatives of science and technology as moral expedients create a powerful *ethos* that may partly explain what occurred" (p. 266). Katz illustrates this point by rhetorically analyzing a technical memo written in June 1942 by a Nazi writer attempting to convince his superior officer "of the necessity for technical improvements to the vans being used in the early Nazi program of exterminating the Jews and other 'undesirables,' just months before the Final Solution of gas chambers and death camps was fully operationalized" (p. 256). As Katz points out, it is chillingly possible for a technical memo to be "an almost perfect document" in terms of efficient, clear communication, and yet remain ethically abominable (p. 256). The problem we immediately recognize when encountering the memo today, which Katz all too easily points out, is that "given the subject matter, we might wish to claim that this memo is *too* technical, *too* logical. The writer shows no concern that the purpose of his memo is the modification of vehicles not only to improve efficiency, but also to exterminate people" (p. 257). For instance, witness the author's chilling use of the passive voice to describe a prospective modification to the extermination vans:

> Lights could be eliminated, since they apparently are never used. However, it has been observed that when the doors are shut, the load always presses hard against them as soon as darkness sets in. This is because the load naturally rushes toward the light when darkness sets in, which makes closing the doors difficult. Also, because of the alarming nature of darkness, screaming always occurs when the doors are closed. It would therefore be useful to light the lamp before and during the first moments of the operation. (as in Katz, 1992, p. 256)

Here, human beings are referred to as "the load," which panics not in response to demented, state-sponsored extermination, but "the alarming nature of darkness." Clearly the Holocaust and this memo are extreme examples of the "ethic of expediency," but Katz goes on to show how "it is the ethic of expediency that enables deliberative rhetoric and gives impulse to most of our actions in technological capitalism as well" (p. 258). These concerns extend to "technical communication and deliberative rhetoric, and by extension writing pedagogy and practice based on it" (p. 258), because they are deeply engrained in Western rhetorical practice. In fact "In the *Rhetoric* Aristotle…seems to collapse all ethical questions in deliberative discourse"—that is to say, discourse designed to deliberate the future ramifications of a question or problem, and to debate means of possible implementation—"into a question of expediency" (p. 260).

What does the "ethic of expediency" have to do with engineering pedagogy, and more specifically with creativity in engineering? In fact the problem with ethical engagements in engineering pedagogy—and a possible explanation for both Cech's "culture of disengagement" and Gambetta et al.'s jihadist engineers—is that they, too, have the tendency to overemphasize expediency. "[W]e are in the habit of giving expediency too much free reign" (p. 261), Katz reminds us. This habit persists, and it dictates our pedagogy, as well as our assessment of that pedagogy, at our peril. Although Katz would be the first to point out that the role of the "ethic of expediency" in the Holocaust played out in a certain place and time, a similar ethic underwrites technological progress and communication in the United States and throughout most of the Western world. Whereas the "communal *telos*" or possible end-goal of life in Nazi Germany was "the power of the State itself, as embodied in technology, Party, and Führer," in the United States it is, according to Katz, "economic progress" (p. 270). The danger for us may not be the rise of an iron-fisted state and the imminent extermination of undesirables, but there is the possibility that "technological expediency in the guise of free enterprise can become de facto both a means and an end" (p. 270). This possibility becomes compounded for engineers, the work of whom Katz thinks of explicitly at the end of his essay:

> For in an age when it is sometimes considered "economically rational" to accept high insurance costs on plane crashes rather than improve the safety of planes; when Ford Motor company decided that it would be more cost-effective to incur the law suits (and loss of life) caused by the placement of the gas tank on the Pintos rather than fix the problem, and only changed its mind when an equally expedient solution was found; when *person*nel are now referred to as Human Resources, like shale or oil, with the metaphorical implications that they (we) can be used up and disposed of or replaced when need be…—when every field strives to be scientific and technical and decisions are made and consequences weighed an

> valued argued on the ethic of expediency only—the holocaust [sic] may have something to teach those of us in technical communication, composition, and rhetoric. (pp. 272–273)

The "ethic of expediency" is not the only historical factor that caused the Holocaust. But the Holocaust, as well as many other human-caused ethical disasters, including those Katz lists here are all outgrowths of a culture in which the "ethic of expediency" has immense rhetorical power. Firsthand encounters with the "ethic of expediency"—as a general rhetorical theory, and in its manifestations such as the Nazi memo Katz describes—have a great deal to teach engineering students. While it might not rectify Cech's "culture of disengagement," if treated as a pillar of engineering education it could have a meaningful (if incremental) impact on engineering ethics by acknowledging the need for an epistemological shift.

Supplementary Ethics: The Case of *A Guide to Writing as an Engineer*

At UCSB, where I teach in a dedicated Engineering Writing Sequence (or "E-Sequence") administered by an independent Writing Program,[2] most if not all writing instructors use David F. Beer and David A. McMurrey's *A Guide to Writing as an Engineer* (4th ed.) in their lower-division engineering writing courses. The proven utility of this textbook as a tool for teaching technical communication is a key reason why evaluating its presentation of ethics has relevance here. Throughout the textbook, Beer and McMurrey offer engineering writing students a concise and often compelling guide for how to write as an engineer, which speaks in the language of engineering. Against general technical writing manuals, they claim their book "stays close to the real concerns engineers and engineering students have in their everyday working lives" (p. vii), and in most ways they succeed. The chapter on grammar and stylistics, for instance, titled "Eliminating Sporadic Noise in Engineering Writing," is pragmatic, crystal clear, and even interesting (no easy task

[2] In collaboration with the UCSB College of Engineering, the UCSB Writing Program offers engineering students a series of discussion- and workshop-based writing courses: Writing 1E: Approaches to Writing for Engineers, Writing 2E: Academic Writing for Engineers, and Writing 50E: Writing and the Research Process for Engineers (students may test out of Writing 1E, depending on their performance on the University of California's system-wide Analytical Writing Placement Exam). All of these courses are capped at 25 students, and all are limited to College of Engineering students in their freshman or sophomore years of study. The courses are staffed by full-time faculty lecturers and professors who teach in the sequence on a regular basis. The sequence mirrors the general first-year writing track available to all students, with a few key differences, including goals stated in the course catalogue to help engineers learn to write in a variety of engineering-related genres, such as technical reports, multimedia presentations, and collaborative projects. The UCSB Course Catalogue (2016) description for Writing 1E explicitly states that the course may "include a consideration of ethics within the world of engineering," and Writing 50E, "the final course in the sequence, prepares students for the profession of engineering through a focus on collaborative research and teamwork."

for a chapter teaching grammar to engineers!) (pp. 12–39). So is the chapter on larger organizational concerns, and strategies for making ideas accessible and tone-appropriate for nontechnical audiences ("Guidelines for Writing Noise-Free Engineering Documents") (pp. 40–71). Both chapters, and indeed the book at large, are popular with engineering students, who, the authors point out, may "dislike writing. After all, don't you go into engineering because you want to work with machines, instruments, and numbers rather than words?" (p. 1). Throughout most of the textbook, Beer and McMurrey not only describe but demonstrate noise-free, accessible writing. They even take on an activist role periodically, especially when they encourage engineers to get away from usage of the passive voice as much as possible in their writing, which is something of a controversial statement given that some engineers still feel that the passive voice feels more appropriate in technical contexts.[3]

Not surprisingly, the Beer and McMurrey textbook contains a chapter on ethics, "Chapter 11: Ethics and Documentation in Engineering Writing." When I first began teaching in the E-Sequence, I considered assigning this chapter in Writing 2E as a basic introduction to engineering ethics, which would have supported the "Technology and Society" unit I teach regularly (I describe this unit in depth below). In this context, I rapidly discovered the chapter's limitations. Here, the textbook's exceptional clarity, as well as its utility and impact, unexpectedly stalls, and can actually prove disruptive in the classroom because it evades rather than confronts the ethical considerations it raises. I offer its structure and content as worth considering because it represents in microcosm how discussions of ethics frequently play out—or, more commonly, fail to play out—in engineering contexts.

Beer and McMurrey offer their discussion of engineering ethics under the sign of two ambitious epigraphs. One is from an ethics textbook by Mike W. Martin and Roland Schinzinger, *Ethics in Engineering* (4th ed.), which reminds students,

> Technology has a pervasive and profound effect on the contemporary world, and engineers play a central role in all aspects of technological developments. In order to hold paramount the safety, health, and welfare of the public, engineers must be morally committed and equipped to grapple with ethical dilemmas they confront. (as cited in Beer et al., p. 229)

The other is by Marcus Aurelius: "If it is not right do not do it; if it is not true do not say it" (as cited in Beer et al., p. 229). Both epigraphs set the stage for what we might expect to be a serious ethical encounter. Clearly, large-scale ethical questions are within the arena of the student-reader's expectation. Based on these epigraphs, it is not difficult to imagine the textbook addressing questions of "right" and "wrong," "truth" and "lie," and offering case studies or examples of considerable ethical complexity involving "the safety, health, and welfare of the public."

Yet after a brief, vague introduction to the relevance of ethics for engineers—who are, the authors adamantly remind their student reader, citizens of the world and "no different from anyone else" (p. 229)—Beer and McMurray suddenly defer their ethical presentation altogether. Instead, in a bold text block at the top of the

---

[3] "Nowadays engineering writers are getting away from the rigid use of the passive. Sentences become more vigorous, direct, and efficient in the active form. By showing that a *person* is involved in the work, you are doing no more than admitting reality" (p. 57)

second page of the chapter, they provide the following "Note," which is really a disclaimer:

> The material on engineering ethics contained in this section is only a general introduction. Much more could be said on the topic, and many books and websites go far more extensively into what we touch on here. Workshops, courses, and distance learning are also available from many sources. An excellent starting place for those who wish to go deeper into this vital part of an engineering career is Texas Tech University's Murdough Center for Engineering Professionalism at www.niee.org. (p. 230)

While one can understand the impulse to refer the reader to more comprehensive resources on engineering ethics, in this textbook that otherwise speaks so authoritatively and convincingly to engineering students, even occasionally taking on an activist role insofar as it risks offering opinions on how engineers *should* write, the move is unexpected. This inaugurates a pattern of understatement and evasion, and what proves to be a very limited discussion of ethics. Already working in a space where ethics have been compartmentalized into a single, late chapter, the authors now restrict their explicit advice on ethics to "Five Communication Concerns" (p. 230). These are primarily related to the business and legal sides of engineering: copyright infringement, tampering with results, withholding adverse information, writing unclear instructions, and omitting safety warnings (pp. 230–232). Notice that none of these "concerns" even broaches larger questions of human technological development, let alone anything like Katz's "ethic of expediency"—except insofar as the chapter itself constitutes an example of a far too expedient approach to ethics. They also provide, as "tools for ethical decision-making," two established ethical codes, one by ABET, the other by the Institute of Electrical and Electronics Engineers (IEEE). "With such documents in hand," they write, "you can refer to guidelines that will support your decisions to hold out for strictly ethical writing (and other activities) as an engineer" (p. 232). Unfortunately it is hardly clear how having "such documents in hand" in a true ethical bind—where corporate, governmental, and/or academic power hierarchies are very likely in play—would provide a means to help make one's case for oneself. Note, too, how these "other activities," which presumably constitute the *real* work of engineering, are parenthetical here. The rest of the chapter deals entirely with the ethics of documentation and a guide to IEEE citation and reference style.

"If it is not right do not do it; if it is not true do not say it": Beer and McMurrey presumably recognize the importance of helping engineers to learn how to draw these lines described by Marcus Aurelius, at least insofar as they include a dedicated chapter on ethics in their textbook. Yet their conspicuously expedient chapter focuses almost entirely on rudimentary business ethics and evades engaging serious ethical discussions, all under the guise of focusing on writing and communication issues. Compounding this evasion, the textbook offers a sprinkling of provocations reminding students of the importance to think about urgent engineering ethics controversies, mainly related to current issues connected to the environment and sustainability. One feature of the fourth edition of the Beer and McMurrey textbook is the addition of what the volume calls "tech boxes," which are distinct, separated "text boxes that briefly describe exciting innovations and advances in the field of engineering" (p. vi). These boxes, scattered throughout the volume every 8–12 pages, are a textbook convention, and in this specific instance possibly the innovation of the publisher rather than these two authors. The boxes are shaded light gray, deploy an alternate sans serif font, and appear raised above the surface of the regular textbook. They are shaded on the left and bottom to give the visual impression of a box lifted off the page and casting a shadow. Throughout most of the textbook, the "tech boxes" serve the primary function of providing light distraction for the reader as reminders of the technological innovation that is ostensibly the domain of the engineer. They often refer to optimistic tidbits about new energy technology: "Knee power: Engineers at several British universities have developed a wearable piezoelectric device that converts knee movement into electricity, which could in turn be used to power gadgets such as heart rate monitors, pedometers, accelerometers, and mobile phones" (p. 87); or "Jellyfish power: Researchers at Virginia Tech have designed a silicone underwater robot that moves like a jellyfish and that is powered by the water around it enabling it to operate apparently forever" (p. 47). They make the volume hip, politically sensitive to sustainability, and assuredly guarantee the need to issue a fifth edition as soon as the novelty of the particular innovations mentioned expires.

Additionally, they advertise the textbook's companion website. All the tech boxes contain instructions, "For details, see the Preface for the URL" (p. 47), although these sources are hardly easy to track down (to do so involves following a series of embedded links). But in the ethics chapter, where there are four tech boxes—a greater density than anywhere else in the volume—they take on a different function: they serve as reminders of the ethical controversies facing human society as it advances technologically. Rather than interesting tidbits, these tech boxes contain provocations posing serious ethical questions:

> Sonar technology deafening cetaceans?: The UK's Whale and Dolphin Conservation Society alleges that low-frequency military sonar technology used in the search for oil and gas is causing hearing loss in cetaceans (whales, dolphins, and porpoises), injuring them and causing them to strand themselves. The February 2011 mass stranding of over one hundred whales on the coast of New Zealand is seen evidence [sic] (p. 230)

The other three questions include "Fracking causing pollution?" (p. 232), "Drone aircraft violating privacy?" (p. 239), and "Keystone pipeline to pollute the Great Plains?" (p. 242). When one follows the instructions to examine the textbook's companion website for details, there are links to external sites that engage these various issues.

Beer and McMurrey's "tech boxes" illustrate the fragmentation of engineering ethics pedagogy, especially as it relates to the teaching of writing for engineers. In their paradoxical attempt to at once intrigue the reader with ethical questions and at the same time to disclaim their authority and liability, to provide parenthetical and perhaps even lighthearted concerns visually separated from the serious technical business of engineering writing and business ethics, they reflect the frequent shallowness of ethical discussions in engineering contexts. Beer and McMurrey, and the textbook's publisher Wiley, clearly recognize the increasing institutional pressure to touch on ethics in this textbook, which, although first published in 1996, did not contain dedicated ethics content until the second edition (2004).[4] Yet by compartmentalizing these ethical concerns in a limited chapter, and furthermore separating urgent technological ethics problems from relatively mundane (and largely intuitive) guidelines for approaching what are essentially business ethics, they miss opportunities to explore the complexity of engineering ethics as well as opportunities for teaching engineers how to think creatively. At the transitional moment in the chapter, when it shifts from ethical concerns as such to the documentation section, Beer and McMurrey leave their reader with a relatively unconvincing hope for the impact of their ethics content:

> Whatever your personal outlook, it's worth remembering that the study of ethics will not necessarily make you a 'better' person, but it will make you a more knowledgeable person when you come face to face with difficult professional decisions. We hope that this section has given you some insight and tools that will allow you to be an ethical researcher, writer, and engineer. (p. 233–4)

The authors' confusing remark that it is not the business of ethics study to make the engineering student into a "'better' person" notwithstanding, to simply claim that ethical training is meant to culminate in abstract, amoral "knowledge" reflects a limited view of how ethics can and should be approached in engineering contexts. The ending of this section confirms the textbook's discomfort with ethics. Despite its originality, classroom utility, and considerable appeal to many engineering students, it ultimately evades approaching ethical topics in any meaningful depth.

Engineering Ethics Pedagogies and a "Culture of Disengagement"

The Beer and McMurrey textbook limits its engagement of ethics as a matter of structure and content. But the failure to come up with a meaningful ethical engagement for student engineers lies less on the shoulders of these two authors than on the discipline at large, which itself is an expression of widely held cultural assumptions and practices. As Katz (1992) illustrates in "The Ethic of Expediency," engineering communication—and perhaps all technical communication—shirks ethical responsibility because it participates in a culture where expediency, in the direction of economic success and technological advancement, can be taken for granted as an integrated *telos* (p. 270). Despite considerable pedagogical planning and vigilance to make certain care is taken to promote meaningful engineering ethics education, as we shall see, the factors delineated below can be exceptionally hard to reckon with. Few twenty-first century engineers in the Western world would probably admit a willingness to enter Faustian pacts with indefensibly harmful projects and

[4] The 2004 edition added a chapter titled "Documentation and Ethics in Engineering Writing." In the 2009 third edition, the present order was introduced, where the discussion of general ethics occurs first and the documentation material second: "Ethics and Documentation in Engineering Writing."

technologies. The goal of engineering ethics education should be to remind them that, under the influence of mad political regimes and during historical emergencies—and even accidentally—engineers have all too frequently entered into such alliances.

There are at least three distinct reasons why ethics often constitute an abstract predicament for engineering education, rather than an arena for serious philosophical debate or creative opportunity. All of these reasons should be brought to the attention of engineers during their undergraduate experience as key pillars of their ethical training. In the first place, there is the generally observable fact that the Hippocratic threshold of "doing no harm" is murkier in technological development than in other fields, such as medicine, or human subjects research in all fields. In a competitive technological marketplace, researching the long-term societal impacts of a new technology—for instance, a technology with the power to change human social interactions, such as the currently stalled Google Glass project (Bilton, 2015)—can sometimes seem secondary to simply developing a functional and marketable prototype. The question of causality, or what the philosopher Martin Heidegger (1954/1977) called "the instrumental conception of technology," looms over all technological advancement, which is never merely a "means to an end" but a force with the capacity to restructure human interactions and to make possible new ways of knowing and being (pp. 5–6). Yet while some specific impacts may be predicted in advance, whether a given technology's impact will be ultimately beneficial or harmful, and for how long, is hard to determine. In the case of Google Glass, how could one possibly weigh, for instance, the benefit of readily available information in one's visual field against privacy concerns? Should such a technology become widespread, it would moreover enter the technological marketplace as part of an ecology, where the implementation of any given new advancement always interacts with and changes existing technologies and practices surrounding their use. The ethical question of whether Google Glass is a "good" or a "bad" innovation rapidly becomes secondary to the seemingly more practical questions of working out its feasibility, uses, and marketability.

A second reason why engineering ethics are problematic, which should also be made clear to student engineers, comes from the broad array of work that falls under the blanket term "engineering." By this I do not only mean the various subfields of engineering, although there are certainly specific ethical considerations that are more relevant to mechanical engineers versus, say, computer engineers. There are yet more fundamental differences, and contradictions, between the ethics of technological innovation—with which the present study is primarily concerned—and the ethics of engineering a solution to a localized technical problem. In the Google Glass example cited above, the development of a new technological platform primarily raises ethical questions related to privacy concerns and social change. Engineering a new winglet for the Boeing 737, by contrast, would be attended by a completely different set of ethical considerations, primarily related to questions of safety versus efficiency and sufficient testing. Whereas the ethics of implementing Glass are largely speculative, and thus require considerable imagination to hypothesize adequately, improving winglet design on an airliner would require the observance of established guidelines and care to attend to appropriate testing procedures.

Third, and perhaps the factor of greatest impact on engineering ethics education—and the point that all engineering students should identify early in their academic and professional lives—is that meaningful ethical decisions in engineering technological developments often exist far away from the individual, day-to-day practice of engineers, whether working in academic, corporate, or freelance settings. Although the figure of the self-directed, creative, individual inventor may still loom in the background mythology of engineering, in our time nearly all major technological development is incremental and collaborative. Individual engineers have especially limited agency when it comes to directing their work in ways that advance public welfare (Swierstra & Jelsma, 2006; van de Poel & Verbeek, 2006). While working for a weapons manufacturer implies some level of consent to the firm's agenda, and while an engineer employed in such a firm should certainly be aware of the possible negative impacts of his or her contribution, one's specific role in the development of any particular technology is likely to be partial, fragmented, and abstract. For this reason, theorists of engineering ethics pedagogy frequently prioritize professional ethics, including case studies of behavior, and strictly distinguish such ethics from personal morality: "Engineering ethics is professional ethics, as opposed to personal morality. It sets the standards for professional practice, and is only learned in a professional school or in professional practice" (Harris, Davis, et al., 1996; Harris, Pritchard, et al., 1996, p. 93).

In the context of engineering ethics pedagogy, these three factors—the complexity of predicting the ethical fallout of a given technological innovation, the significant difference in relevant ethical considerations between engineering subfields and objectives (especially technological innovation versus focused problem-solving), and the limited agency possessed by the individual engineer—combine to influence the shape of discussions, readings, and assignments. They force instructors to conflate and to reduce ethical priorities, while usually, at the same time, also forcing them to also be *too* comprehensive in what they include on their engineering ethics syllabi. They moreover limit the ability of researchers to discover the extent to which ethical knowledge has been meaningfully transmitted, whether that is conceived as a cognitive skill, moral courage, or—what is of central interest to the present study—ethical creativity. Only when presented as topics in ethical education rather than as problems do they become essential lessons for student engineers, rather than constraints.

Instructive, here, may be an engineering ethics course at the University of Illinois at Urbana Champaign (UIUC), Electrical and Computer Engineering (ECE) 216: Engineering Ethics, and the study Golnaz Hashemian and Michael C. Loui (2010) conducted to measure student learning. This course illustrates how, when ethics are seen as a loose amalgam of moral problems rather than as an essential domain of engineering—part of engineering's epistemology—fragmentation and shallow abstraction pervade and limit student learning. According to the authors—Loui is the instructor, while Hashemian was his undergraduate assistant researcher whose role was to facilitate unbiased interviews—the goals of the course range widely across ethics as a field:

> The course emphasized professional ethics in engineering, including professional responsibility, conflict of interest, confidentiality, safety and risk, relationships between managers and engineers, whistle-blowing, and codes of ethics. ECE 216 used a textbook…supplementary readings, videos, short scenarios about common problems, and case studies about major disasters. For their research papers, most students chose ethical issues in a current controversy, such as nuclear weapons, stem cell research, genetically modified foods, and copyrights for digital media (p. 203–4).

What is initially striking in this course description is the exceptionally broad range of educational goals present, as well as their apparent bifurcation. On the one hand, the course follows the imperative to limit engineering ethics to professional ethics offered (Harris, Davis, et al. (1996) and Harris, Pritchard, et al. (1996)), some of whom also authored the textbook used by the UIUC course. ECE 216 "emphasized professional ethics," and the list of ethical topics it focuses on, while certainly relevant, are the sort one might see replicated in a business ethics course. Yet it also provides scenarios and "case studies about major disasters" specifically relevant to engineering. Although it is unclear whether students developed their research topics themselves or chose them from predesignated options, their topics—"nuclear weapons, stem cell research, genetically modified foods, and copyrights for digital media"—are notably ambitious, and clearly extend to larger cultural conversations regarding technology and innovation (i.e., these would be topics worthy of a "tech box" in the Beer and McMurrey textbook).

This course appears to be taking a comprehensive approach to engineering ethics, blending the emphasis on "professional ethics" advocated by Harris, Davis, et al. (1996) and Harris, Pritchard, et al. (1996) with a broader vision, at least in the research paper component of the course. When Hashemian and Loui describe their research instrument for determining student ethical knowledge, however, their assumptions and goals clash with this seemingly broad vision, and in terms that closely recreate what we witnessed in the Beer and McMurrey textbook. After the course described above, they presented students (including both students who had completed the course and students who had not, to provide a control sample) with two of the specific case studies used in the ECE 216, both of which were adapted from the textbook by Harris and colleagues. The first describes an "intermittent glitch" in a payroll system, where an engineer must confront several variations of a case in which an automated computer system miscalculates pay by an amount of several pennies, and interact with a third party who changes, in the variants, from external auditor to employee (pp. 204–205). The second deals with an engineer who accidentally discovers that the microwaves built by his company have radiation levels that exceed safety standards (again, there are several variants: "In the first variation, the microwave ovens are manufactured by the engineer's employer…In the second variation, the student encounters an employee who was responsible for measuring radiation levels. In the final variation, the student is no longer a new-hire, but an experienced engineer with a Professional Engineer's license" (p. 205)). While there may be some educational value in exposing students to these two very specific ethical situations, as tests for whether students have actually achieved ethical knowledge, they mainly reflect an awareness of the limits of power and agency within the engineering world, which should be presented as the true lesson here and one that deserves to be approached directly, rather than obliquely through case studies. The findings of the study, in which "[a] random and purposive sampling was used," included both students with no prior ethics education and students who had excelled in ECE 216, are for the most part highly intuitive. ECE 216 students, with training in cognitivist, case study based ethical training "made a clear plan of action" (p. 206) and "recognized moral problems" (p. 207). In the case of the "Payroll" question, however, "differences between the responses of ECE 216 students were also seen… two ECE 216 students felt that the problem was 'not my business'" (p. 208). Not surprisingly, "the students were reluctant to talk about their feelings," but when they did open up, those "who decided to take positive action said they eventually felt better about themselves, more responsible, more confident" (p. 209). These responses—that behaving ethically could make students "feel better about themselves"—are mainly intuitive, especially given that there are few concrete measures for what makes a "good" person. The deference to relatively banal ethical standards in the UIUC study mirrors what we witnessed in the Beer and McMurrey textbook. It, too, evades asking engineers to be responsible for serious ethical concerns.

Academic discussions of engineering ethics education, in short, are largely inscribed in reductive, overly expedient pedagogical inclinations and strategies of self-evaluation that fail to meaningfully engage the serious ethical questions faced by engineers in our time. As Walling (2015) points out, in her essay that takes a legitimate stand against this problem, "Since case studies inherently focus attention on texts, they also may have the tendency to intellectualize and compartmentalize ethical decision making and obstruct efforts to produce ethical behavior in our students" (p. 1649). Productive remedies may include Walling's "moral experimentation" model, where students are encouraged to explore and record ethical encounters and questions in their real lives. But outside the emerging field of engineering ethics pedagogy research, the question of whether ethical education in engineering is even possible threatens to unravel these pedagogical debates, and may extend to the very nature of engineering as a field. Cech (2014) goes so far as to question "whether engineers, particularly neophytes, even consider public welfare concerns to be within the purview of their responsibility as professionals" (p. 45). According to Cech,

> despite institutional enthusiasm for nurturing aspiring engineers' engagement with questions of public welfare, the restructuring of accreditation, and a broad recognition that such engagement is important to the role of engineers, engineering education fosters a culture of disengagement that defines public welfare concerns as tangential to what it means to practice engineering. (p. 45)

In fact, ethical indifference may even increase over the course of a degree in engineering, because of three deeply engrained "ideological pillars" within this "culture of disengagement": an "ideology of depoliticization" in which "nontechnical concerns" are deemed irrelevant to what is perceived as actual, authentic engineering work; the "technical/social dualism, which devalues 'social' competencies such as those related to public welfare"; and the inertia of a "meritocratic ideology," which reifies rather than questions existing social structures (p. 45). Although interventions in ethical education, such as those described by Hashemian et al. and Walling, may have some incremental effect on the ethical sensibilities of engineers, for Cech the deck may be stacked, so to speak, against engineering ethics educators because of fundamental assumptions and practices surrounding the work of engineers: a culture in which ethics are always already supplemental.

The gravity of this situation becomes compounded when one considers the findings of Gambetta and Hertog (2016) in their provocative study *Engineers of Jihad: The Curious Connection of Violent Extremism and Education*, which received widespread media attention in the wake of the March 22, 2016 Brussels terror attacks. Although these authors exercise caution when drawing direct causal links between practices of engineering education and terrorist activation, they recognize the disproportionate number of perpetrators of Islamist violence who have obtained technical training and degrees in engineering and the sciences. According to Gambetta and Hertog, although the number of engineers in the general population of countries of origin for perpetrators of extremist terror is only 1.3 %, in a sample of 231 extremists, 18.7 % had formal training in engineering (pp. 14–15). "The odds of finding an engineer in the sample is *seventeen times greater* than what we would expect if engineers were as likely to radicalize as the male adult population in general" (p. 15). As Gambetta and Hertog explain, this is

> doubly puzzling when set against commonsense expectations. While we readily accept that the dispossessed are natural candidates for extremism, we are at a loss to comprehend why well-off, educated men should join the ranks of jihad. And why would individuals with a technical mind and training in modern technology have a penchant for a movement at once violent, religious, and, in many cases…permeated by antiscientific beliefs? (p. 3)

Why indeed? Speculating on the narratives and psychological conditions that would take an engineering student from working through fluid dynamics problems in the university library to strapping on a suicide belt is a task that lies far outside the bounds of the present study. But an answer to the puzzlement expressed by Gambetta and Hertog might be available in rhetorical theory, which has implications for our consideration of engineering ethics pedagogy and the possibility of making ethical engagements a creative endeavor.

Toward Creative Ethics in Engineering: A Case Study and Model Assignment

The goal of the present project is neither to obsessively criticize existing approaches to the teaching of engineering ethics, nor to present a magic bullet solution to the question of how to better foster ethical development in student engineers. Rather, it is to identify the treatment of ethics as a supplemental concern for engineers as a grave problem, and to propose that direct engagements of ethical problems *as* engineering problems—in particular, the theory of Katz's "ethic of expediency" as well as instances of this rhetorical phenomenon—may be an alternative strategy.

The UCSB Writing Program's Engineering Writing Sequence provides a unique setting for exploring such alternative strategies, and I took a first step in this direction in a case study I performed in the section of Writing 2E: Academic Writing for Engineers I taught in Winter Quarter 2016. This course marked the second time I had taught Writing 2E, and the fifth quarter I had taught courses for the Engineering Writing Sequence (although I have taught other courses at UCSB and elsewhere in technical writing and communication, including writing for the health professions and social sciences). IRB approval was obtained for this study, as was permission to cite from student work. The course had 25 students, all of whom were enrolled in UCSB's College of Engineering in their first 2 years of study. The course required students to complete three major writing projects: a rhetorical analysis of an engineering-related genre, a technical recommendation report offering a solution to real-world problem, and a dedicated ethics engagement in what I called a "Technology and Society Essay." Ethical concerns were on the table for us all quarter long. On the first day of the term, we watched an episode from the satirical British science-fiction television series *Black Mirror* titled "The Entire History of You" (2011), which depicts a near-future world where implantable artificial memory chips have become widely available, turning one's daily experience into recordable video and audio possible to be summoned on demand. The questions raised by this series—especially the question of the extent to which technologies shape human behavior and societal standards and interactions—set the stage for a number of conversations about technologies and their unexpected ramifications.

Obviously Katz has a large role in the present article, and it should be no surprise that I routinely teach his theory to my students in E-Sequence courses early in the quarter. On the first day, I also showed students an adapted version of the Nazi engineering memo mentioned by Katz in his article, which the article reproduces in its entirety. Eliminating identifying characteristics (i.e., the place and date of Berlin, June 5, 1942, and the note "Submitted for decision to *Gruppenleiter* II D, *SS-Obersturmbannführer* Walter Rauff" (as in Katz, 1992, p. 256)), I gave students, working in small groups, 10 min to attempt to imagine the rhetorical situation. I asked them questions including *who wrote this letter and to whom? And to deliberate the rectification of what problem?* Although I was able to overhear students murmuring that they suspected this might be a van designed for the extermination or transport of animal cargo, and I heard at least one student question whether this possibly a Holocaust setting, none explicitly ventured that this might be a Nazi memo when I asked for hypotheses once we reconvened as a large group.

Though we had no easy solutions, ethics and technological development were on our radar beginning on the first day of the quarter, even as we spent the largest segment of our time together working on a highly technical recommendation report. In the final Technology and Society Essay, however, I sought to provide students an explicit encounter with the ethical controversies surrounding new technologies as an *opportunity* for creativity and learning. In this assignment, students were asked not simply to investigate the potential controversies surrounding a given technology, but to research an obviously problematic technology and find the means to *defend* it ethically. Here is the exact language of the prompt, which in class I sometimes referred to as a "devil's advocate" paper:

> Write a *3-4 page paper* that argues convincingly *in favor* of a potentially problematic technology, invention, product, or software platform. First, choose a technology and do some research, identifying why the technology is problematic as well as the specific arguments surrounding its existence and/or implementation. Then, take an informed, positive position within the larger debate you have investigated, and write a paper that (1) informs the reader of the general stakes of the controversy, and (2) systematically and convincingly persuades the reader to accept your position as a contingent truth.
>
> Regardless of the topic you choose, your thesis must be both *sophisticated* and *debatable* (we'll talk about what this means in class). You may choose any potentially problematic technology you wish, with one rule: you cannot write in favor of a weapon, or any other device that is designed to inflict harm on another human being. The technology you write on may be speculative—that is to say, still in an experimental stage—as long as you account for this in your essay.

I then provided a list of possibilities, including a broad range of technologies (the ideas were widely varied, and included such topics as Google Glass, hydraulic fracking, autopilot, instant replay in sports, genetic engineering, and hydrodynamic bathing suits). The limitation on weapons emerged from my desire to set a basic standard that to be a part of an engineering project explicitly designed to inflict injury is inherently an ethically compromised position, although I made it clear in class that I did not necessarily mean to argue that participation in weapons development is always indefensible.

In this assignment, I sought to offer students a chance to work directly in the rhetorical mode Katz identifies (after Aristotle) as deliberative, by creating an argument in defense of a technology or practice that might have significant negative social or environmental consequences. While the responses were mixed in terms of depth and quality, several students engaged arguments that I believe illustrate how *creative approaches to solving problems of technological development* and an *acute awareness of expediency as an ethical problem* are hardly antithetical. For instance, one student took on the FBI's attempt to force Apple to unlock the iPhone of Syed Rizwan Farook, one of the key perpetrators of the December 2015 terror attacks in San Bernardino, California (the paper was written before the FBI hacked the phone independently). After illustrating the various relevant controversies of the technology of the unlockable smartphone, the student outlines a possible compromise:

> This debate is a very slippery slope; it can have the potential to be an invasion of privacy for hundreds but it can also prevent future attacks similar to the one that took place in San Bernardino. This is why a clear compromise should be made between Apple and the FBI with direct and clear guidelines regarding the current and future use of this technology. The fact that Apple is a very popular company makes this case and any other future cases regarding this technology a public matter. This helps to keep the guidelines clear because of the fact that the consumers and citizens can be seen as an audience. Apple and the FBI must make sure they keep things professional because of their reputations. This ensures that both companies will feel obligated to stay loyal to the agreed guidelines which ultimately makes this technology an asset to the people.

Notice how this *defense* of a highly controversial technological step—in this case, a potential shift in smartphone security policy at Apple—makes its argument with equal attention to the "reputations" of these companies, and their consumers ("the people") for whom this technology must be an "asset" and who "can be seen as an audience." This student clearly recognizes technological policy as implicitly rhetorical, and attends with considerable attention to the ethical dimensions of this problem. Notice, too, the creativity and foresight of this solution. This proposal is attentive to the future relationships between both of these entities—Apple and the FBI—and the publics they serve and profit from (whether through revenue or taxes), although surely these organizations remain primarily devoted to, respectively, turning a profit and maintaining power, authority, and credibility. The argument, ultimately, may reify the "ethic of expediency," but the subtlety of the position taken here suggests some degree of learning has taken place on the question of how to navigate a fraught ethical position.

Another auspicious paper defended smartphone technologies in light of their potentially alienating impacts on human social interaction and mental health. On the one hand, the student acknowledges how cellular phones can disrupt social interactions, and even constitute a form of memory outsourcing: "Cell phones… have an impact on psychology and alter how much information people actually retain. With cell phone data, one can search up anything with the click of a button. This convenient access to any information has changed our attention span and minimized the amount of information people actually absorb." Yet the same student ultimately justifies cellular phone technology in general by comparing it to how Bluetooth earsets appeared as recently as 5 or 6 years ago:

> A device that was widely used just a few years ago and now is not seen around as much is Bluetooth [i.e., wireless Bluetooth earsets]. It caused people to be rude with no etiquette for how to use them. Bluetooth was experiencing a transitional state where people would walk around and talk with it at inappropriate times and use it during times where it could come off as extremely rude. Similarly, phones are experiencing a transitional stage where one sees them everywhere at inappropriate times. However, in a few years from now they will not be as prevalent and people will develop an etiquette for how to use them.

While some minor grammatical and stylistic flaws mark this response, the idea that cellular phones themselves might be in a transitional stage in terms of their socially accepted uses arrives as nothing short of a creative epiphany. This response is perhaps one or two degrees too optimistic, yet it proposes a creative solution to a problem of technological development by imagining, quite on its own, that technologies—even technologies such as smartphones, which many people have unselfconsciously taken for granted—can exist in transitional phases.

A further salient example of how students engage and rethink the logic of deliberative rhetoric in response to the "devil's advocate" prompt occurred in a paper that broke the rules of the assignment. This student defended animal testing in research, which, although animal testing is not a technology per se, I permitted because it raises many of the same ethical questions. The student proposed an exceedingly precarious argument about the relative ethics of killing animals for food consumption versus the advancement of knowledge:

> it is hypocritical for people to argue against animal research if they buy meat, dairy, or any food associated with industrial farms. "Over 99 % of farm animals in the US are raised in factory farms," says the American Society for the Prevention of Cruelty to Animals, "which focus on profit and efficiency at the expense of animal welfare" ("Factory Farms"). In other

> words, most of the animals in the US are raised in farm factories that do not care for them. Their only worry is making the most amount of money from the animals, even if it means that the animals have limited space to move, are unable to engage in natural behaviors, or suffer from depression when separated from their young. Animal research kills animals but in return receives data that can be used on animals and humans. Factory farms, in contrast, kill animals to feed humans which has no benefits to the animals. From this we can conclude that buying meat, dairy or food associated with factory farms is far worse than animal research.
>
> Alternatively, people who do not buy products from factory farms can argue that because we live in an abiding meat-eating world, there is no reason to conduct research on animals when factory farms are greatly contributing to global warming and to the destruction of our world.

In this response, the student—prompted to hypothetically defend animal testing—in fact draws a productive, hardly intuitive ethical line: in a meat-eating world, animal research, which potentially has lasting benefits for both animals and humans, remains ethically justifiable. By entering the realm of deliberative rhetoric through technological controversy, students have the opportunity to explore rhetorics outside the ethic of expediency, which dominates the engineering realm. They also encounter opportunities to generate creative arguments, which, if occasionally unexpected (and provisional), force them to make the positive ethical compromises expedient discourse too easily elides.

Conclusion

At the end of Writing 2E, Winter 2016, I offered students a small amount of extra credit (equivalent to approximately 0.5 % of their final course grade) in exchange for completing an optional questionnaire about their views on ethics and engineering (students were also offered the option to complete an alternate metacognitive exercise for the same extra credit compensation). The questionnaire consisted of the following three questions:

1. When designing project ideas in Engineering Sequence writing courses (Writing 1E, 2E, and/or 50E), what technologies or fields, if any, have felt out of bounds as potential topics? Why?
2. In your opinion, what does creativity look like in your field of engineering?
3. What activities, discussions, and readings related to engineering ethics and/or creativity have been especially beneficial to your intellectual development as an engineer in Writing Program Courses?

Ten out of 25 students responded. Although the responses to Question #1 generally repeated the course requirement that weapons were out of bounds, three respondents remarked that technologies related to medicine or health seemed underrepresented in the class ("I felt like many students chose to do electronic technologies where medicinal technologies were chosen considerably less"). Question #3, similarly, generated mainly unoriginal findings. Question #2, however, generated a set of definitions of engineering creativity that strongly resonate what one may expect from students working in a field so closely associated with narrowly focused expediency. Most of these responses spoke of innovation and especially problem-solving. Here is an inventory of several remarkable responses:

Student #1: As an engineer, you might not be the first person to create something, an engineer's creativity can be used to perfect anything.
Student #2: Something that shows creativity for computer engineers is solving a problem, for example a software or coding problem to fix a program or make it better.
Student #3: My field is computer science. Creativity is usually the development of a new algorithm, or great applications for theories.
Student #4: Creativity in engineering is finding ways to problem-solve obstacles.
Student #5: When engineers encounter problems with no fixed answers, each individual's creativity offers different solutions and finally provides various solutions to one problem.
Student #6: To be creative a person needs to constantly seek a better and more efficient way to reach that goal.

While more than one student expressed that, ideally, creative engineers might be able to "change the world," it is worth noting that in response to a research instrument titled "Ethics in Engineering Writing Questionnaire"—presented at the conclusion of a course in engineering writing that foregrounded ethics—ethics *never* came up as a term to think about alongside creativity and ingenuity.

Examining and practicing subtle, deliberative arguments in favor of potentially problematic technologies in the context of ethics and writing courses, at the very least, provides an encounter with the possibility the ethical priorities might be a medium for creative thinking in engineering. In a world where engineers work, and deliberate, while thoroughly inscribed within technological capitalism's "ethic of expediency," and as part of Cech's "culture of detachment," they should be exposed as often as possible during their education to the potential consequences of imagining that their innovations—and, just as importantly, the way they talk about them—are somehow beyond or outside the realm of ethics. There may be inertia against asking students to carry with them, throughout their academic and professional careers, what might be called an activist ethical sensibility, where engineers no longer can defer to problem-solving, or, as one student put it, the need to "constantly seek a better and more efficient way to reach [a given engineering] goal" as the engineer's primary professional activity (i.e., to the "how" of technological development, unattended by the "why"). Exposure to the reductive logic of overly expedient, deliberative rhetoric, whether in the form of a preposterously unethical technical memo by the Nazis, or by experimenting with advocating for a problematic technology—as well as confronting in class discussion and metacognitive exercises the fraught relationship between engineering and ethics itself—nevertheless imparts an incremental, positive effect on the ethical sensibilities of engineering undergraduates. Only by presenting ethics as an essential domain of engineering, as a core part of the field's epistemology and a pathway to creativity, can we expect engineers to cultivate a meaningful ethical sensibility and to become better educated, more humane engineers.

References

Adorno, T.W. (1997). Cultural criticism and society. In *Prisms* (9th ed., pp. 17–34). (S. Weber & S. Weber, Trans.). Cambridge, MA: MIT Press. (Original work published 1949).

Beer, D. F., & McMurrey, D. A. (2014). *A guide to writing as an engineer* (4th ed.). Hoboken, NJ: Wiley.

Bilton, N. (2015, February 4). Why Google Glass broke. *The New York Times*. Retrieved from http://www.nytimes.com/2015/02/05/style/why-google-glass-broke.html?_r=0

Cech, E. A. (2014). Culture of disengagement in engineering education? *Science, Technology, & Human Values, 39*(1), 42–72.

Gambetta, D., & Hertog, F. (2016). *Engineers of jihad: The curious connection between violent extremism and education*. Princeton, NJ: Princeton University Press.

Harris, C. E., Jr., Davis, M., Pritchard, M. S., & Rabins, M. J. (1996). Engineering ethics: What? Why? How? And when? *Journal of Engineering Education, 85*, 93–96.

Harris, C. E., Pritchard, M. S., & Rabins, M. J. (1996). *Engineering ethics: Concepts and cases* (2nd ed.). Belmont, CA: Wadsworth/Thompson Learning.

Hashemian, G., & Loui, M. C. (2010). Can instruction in engineering ethics change students' feelings about professional responsibility? *Science and Engineering Ethics, 16*(1), 201–215.

Heidegger, M. (1977). The question concerning technology. In *The question concerning technology and other essays* (pp. 3–35). (W. Lovitt, Trans.). New York: Harper and Row. (Originalwork published 1954).

Herkert, J. R. (2000). Engineering ethics education in the USA: Content, pedagogy, and curriculum. *European Journal of Engineering Education, 25*(4), 303–313.

Katz, S. B. (1992). The ethic of expediency: Classical rhetoric, technology, and the holocaust. *College English, 54*(3), 255–275.

Mitcham, C., & Duvall, R. S. (2000). *Engineer's toolkit: A first course in engineering*. Upper Saddle River, NJ: Prentice Hall.

Swierstra, T., & Jelsma, J. (2006). Responsibility without moralism in technoscientific design practice. *Science, Technology, & Human Values, 31*(3), 309–332.

UCSB Writing Program. Undergraduate course descriptions. (2016). Retrieved from http://www.writing.ucsb.edu/academics/undergrad/courses

van de Poel, I., & Verbeek, P. P. (2006). Ethics and engineering design. *Science, Technology, & Human Values, 31*(3), 223–236.

Walling, O. (2015). Beyond ethical frameworks: Using moral experimentation in the engineering ethics classroom. *Science and Engineering Ethics, 21*(6), 1637–1656.

Creative Ways of Knowing and the Future of Engineering Education

Cassandra Groen, Christopher Gewirtz, Adetoun Taiwo, Lindy Cranwell, and Rabih Younes

Introduction

In the earlier chapters of this text, members of the engineering education community have described multiple pedagogical strategies and approaches for enhancing student learning through creative ways of knowing and teaching. There is a unique value in understanding how these educational shifts may affect our graduate students—the future of engineering education—as they embark on their careers. These students form the frontier of engineering education research and provide a perspective on creativity that may be unreachable from other positions in the engineering education community. In this chapter, we shed light on this perspective through a collection of reflections from five students in an engineering education graduate-level practicum course. Students in this course are embarking on their teaching careers and typically take this course during their first teaching experience. As two assignments in this course, students were given the opportunity to review and write a scholarly critique for a chapter of this text and provide an end-of-semester reflection on their evolving views of creativity in engineering teaching based on these chapters and existing literature. While the critique assignment exposed students to cutting-edge, emerging techniques for engineering teaching, the reflection assignment encouraged students to actively engage in the formation of and reflect on their own practices of teaching engineering as informed by research. In particular, the reflections captured students'

C. Groen (✉) • C. Gewirtz • A. Taiwo
Department of Engineering Education, Virginia Tech, Blacksburg, VA, USA
e-mail: cgroen@vt.edu

L. Cranwell
Department of Civil and Environmental Engineering, Virginia Tech, Blacksburg, VA, USA

R. Younes
Department of Electrical and Computer Engineering, Virginia Tech, Blacksburg, VA, USA

D. Bairaktarova, M. Eodice (eds.), *Creative Ways of Knowing in Engineering*,
DOI 10.1007/978-3-319-49352-7_9

own academic backgrounds, overall opinions of the role of creativity in engineering teaching and learning, and the ways in which their own educational practices have evolved as they gained a greater understanding of this topic.

Through the voices of these students, this chapter serves as a representation of our view regarding the future of creative ways of knowing in engineering education. We hope that this chapter serves as an inspiration for the continued development of engineering teaching and learning through creativity and drawing on many ways of knowing.

In the following sections, we present actual reflections from graduate students at Virginia Tech, who are preparing for careers in engineering education, as they discuss their experiences and thoughts for the future.

Cassandra Groen

Cassandra's initial interest in engineering started when she was in high school as her family encouraged her to combine her talents in math and science with her love of structural design and desire for a broader impact on society. She graduated with Bachelor's and Master's degrees in Civil Engineering in 2009 and 2011, respectively. During her graduate work as a Master's student, Cassandra realized her passion for working with other students and conducting education research. She is currently a Ph.D. candidate in the Department of Engineering Education at Virginia Tech where her research interest focuses on professional identity formation in undergraduate civil engineering students.

Initially, education researchers perceived creativity to be an innate characteristic or personal trait that was possessed by specific individuals (Collard & Looney, 2014), inherently communicating that only *certain types of people* are creative. However, this perception is shifting as educators across the disciplines are encouraged to teach creativity by teaching creatively (Collard & Looney, 2014; Craft, 2003). These creative ways of teaching are aimed at fostering students as individual people who are complete with their own sets of prior knowledge, identities, histories, and cultural upbringings. By challenging these characteristics through the encouragement of innovation in a variety of ways and contexts, instructors inherently foster students' creative capacities for learning and knowing as they adjust to and cope with new information (Collard & Looney, 2014). However, I was not a student who was exposed to this inclusive pedagogy and conception of creativity. In my reflection, I explain how I lost my connection to the civil engineering discipline as struggled to learn creatively and practice my creativity during my undergraduate career.

I would say that my experience as an undergraduate civil engineering student was a fairly typical one. Upon entering college, I realized that college life was not like those found in the movies where students sat outside in the quad and philosophized about contemporary issues. It also wasn't like the brochures I had seen that featured students building a mini Indy racer in the lab or constructing a bridge—at least not until senior year. Rather, the majority of my undergraduate experience was laden with long hours of lecture, completing countless pages of calculations, and meeting tight deadlines. This monotonous routine of going to lecture, copying notes, turning in homework, and taking exams slowly became my life and who I was. I succumbed to the idea that this is what engineering education is and this is how teaching and learning is *supposed* to be.

Statics was the most complex course I experienced during my undergraduate career. In that class, our instructor utilized a lecture-based pedagogical technique. He would stand at the front of the class, transfer his hand-written notes to the board, and go through example problems. I struggled with the speed at which he drew his free body diagrams and scribbled the calculations on the board. Soon, I found myself overwhelmed and disengaged. But that was okay; the math was pretty simple, and I could learn the material on my own after the class as long as I had my notes. To my dismay, my approach did not work, as evidenced by my receiving a "C" as a final grade in the course. While I had met the minimum grade requirement to continue in my program, I was unsure of my engineering competency. This self-doubt prompted me to question my life choice to become a civil engineer and my intelligence. If statics is the "bread and butter" of civil engineering, how was I going to survive the harder courses in my program? Despite my concerns, I kept pushing through my courses and pushing through the math. Creativity was something that was associated with a romanticized view of engineering and my "learning" of concepts dwindled to rote mathematical practice of textbook problems.

Today, as an emerging engineering educator, I draw from my own prior experiences to influence how I approach teaching and engage my students in new and creative ways. As a student, I became objectively disconnected from the course material and my major. Now, as an engineering instructor, I strive to inspire my students and strengthen their disciplinary ties through creative pedagogies.

Specifically, my teaching philosophy and practices are inspired by emerging research within this field, particularly regarding the ways in which instructors bring students' individual creativity and personalities into the classroom. In a thermodynamics course, Diana Bairaktarova (Bairaktarova & Eodice, 2017) introduced a project that prompted students to submit a video of performed written essays, poems, songs, stories, or commercials that explained thermodynamics course. As I reviewed student project submissions, I could tell that the students, while a little uncomfortable in front of a camera, enjoyed the freedom to creatively express their ideas and were proud to bring their own identities and styles into their work. They developed comedies, poems, and metaphors to link course topics such as the Ideal Gas Law, Maxwell's Demon, or the basic laws of thermodynamics to their everyday lives.

Another way in which instructors bring student identities into the classroom is through writing. In an engineering writing course, instructors are beginning to utilize ethical cases as a creative and engaging way to teach technical writing skills to students (*see chapter titled "Ethics and ingenuity: Creative engagements of ethical dilemmas in the engineering writing classroom*"). While writing is not always viewed by students as a necessary professional skill, Moore draws on the conflict of knowledge (Craft, 2003) to create writing prompts that ask students to choose a contemporary technology and present a self-directed "devil's advocate" discussion corresponding to the technology. By using a writing assignment of this nature, Moore is encouraging students to creatively question their own views and perspectives while negotiating existing conflicts on a technology that is of interest or importance to them. By struggling through these conflicts, students will gain an ethical awareness as engineers and enhance their abilities to generate and discuss nontraditional ideas for new engineering problems and solutions. With this assignment, Moore is linking individual student interests and goals to the ethical formation of future engineers through writing—which is something that I would love to try in my classroom with my future students.

As a new engineering educator with one semester of teaching experience, I am inspired by this emerging body of research. It is encouraging as a new educator that many of the ways in which researchers and instructors are incorporating creativity into their classrooms is not as labor-intensive as one may initially assume, particularly when teaching traditionally tough, fundamental engineering courses such as thermodynamics. These new and creative ways of teaching spark students' own creative interests by giving them more agency in directing course projects and incorporating their own identities into them. Creative ways of teaching for creative ways of knowing provide instructors with the tools to make knowledge more accessible, relevant, and applicable to students and their everyday lives. I know that as an instructor, I plan to add this body of work and incorporate many of the topics discussed within this volume into my own classroom practice to keep my students engaged and involved in their learning.

Chris Gewirtz

Chris is a Ph.D. student in the Department of Engineering Education at Virginia Tech. He has always had a knack for breaking and bending rules since he wanted to be a mad scientist when he was young. He has spent one semester teaching the introductory engineering workshop in the Spring of 2016. He thinks engineering has so much potential to change the world; a potential that is often squandered on slight improvements to technology meant for entertainment or violence, and works towards systematic change as a result. His research interests fall under the umbrella of what we unintentionally teach engineers and what we could intentionally teach engineers through service learning, innovation, macro-ethics, social justice, humanitarian engineering, and default mode thinking.

As a future engineering educator, I recognize the importance of creativity, though I do not consider it to be universally appropriate, and I think that uncritical support of education for creativity deserves the skepticism of some authors (Coate & Boulos, 2012; Craft, 2003). That said, I think that the first thing that we can do to move away from a rigid classroom and toward seeing student creativity is to allow for "resilient" creativity. As an instructor, your role may seem rigid, following along with the classroom policies, procedures, and overall curriculum. As some researchers discuss (e.g., Coate & Boulos, 2012; Henderson & Dancy, 2007; Schön, 1983), there may be institutional features that hinder student creativity.

- You may perceive making adaptive changes to your curriculum to be counter to the department culture.
- You may not have the time to plan the kind of change that you think the curriculum needs.
- You may find that students themselves have resistance to activities that require creativity.

- You may feel pressured to "cover" or address material at the expense of new pedagogy.

Ideally, as an instructor you would want full freedom and control, but this is often unrealizable. Luckily, as I hope my writing will indicate, the inertia and rigidity of institutional structure may not pose too great an obstacle for "resilient" creativity.

The metaphor goes like this: the rules and policies of a class are like an unmovable object and resilient creativity acts like an unstoppable force. The unstoppable force of a student's creativity can be encouraged; this is what intrinsic motivation is all about. Raising intrinsic motivation may have its own barriers. Learning and implementing something new is typically anxiety-inducing, and you may not feel equipped to motivate students in a way that is new to you. I suggest Jones' MUSIC model (Jones, 2009) for attempting to generate student creativity.

But even without eliminating an unmovable object or rallying more of an unstoppable force, creativity still has an avenue. Our unmovable object has cracks (often known as loopholes), and this is where your work on creativity can start. The following is, in my opinion, a good example of resilient creativity in an otherwise rigid structure.

During the Fall quarter of my Junior year, I took a class combining theories of material properties and optics called "Optomechanics & Optical Materials" or "Optical Materials" for short. The class was in the style of a traditional lecture; our class of 11 would come in, listen and watch presentations, ask the occasional question, and complete homework problems based on what we picked up in class. There was a midterm and a final, and there was one more assignment determining our grade: a very short design project. We had to select a series of thin films to meet design constraints such as reflectivity and tilt stability. Most student answers were not that different, after all, we only had control over the thickness of the films, the film material, and the number of films. Still, for me there was enough wiggle room, enough of a loophole in the assignment. Instead of educated guesses and following known thin film designs, I wrote a MATLAB program that simulated combinations of thin films, accounted for dispersion, and found the best combination of film. I went above and beyond, as I might do in a future professional position, and the results were excellent and unique. I also found use for the program in research outside of the class.

Structuring learning experiences so that there is more than one pathway to a clearly measurable result can bring about this creativity. In Virginia Tech's introductory engineering class, which I taught for a semester, we implement a few exercises with multiple pathways.

- We have students perform "Product archeology" on a cell phone of their choice, and we ask them to describe the conditions that influence the design of their particular phone.
- We have students set up their own experiments to measure the period of a pendulum using an Arduino Echolocation sensor setup (with some guidance from their instructors, particularly concerning hardware).

There are a few principles for designing these "loopholes."

- If you can change the rules (a.k.a., the "unmovable object"), they can have a huge effect on stomping out or encouraging creativity, and you can start the work of great change via simple experimentation with loopholes.
- Try not to predict from where or from whom creativity will come. Expectations affect students (Baxter Magolda & King, 2012), and your expectations may open opportunities for one person while closing them for another (students might get the impression that "you must be this tall to be creative").
- Find out what the rules are. Some of them are obvious (i.e., students must turn in homework, late homework receives no credit, the report should have the described features to receive full credit, students must present for 8–10 min). Some rules are implicit (e.g., only the instructor commands the attention of the class, every question has one right answer). You can design loopholes for both kinds of rules.

It may be that a lack of loopholes hides the potential for creativity from your students. If you relax requirements of process that accompany some learning activities and allow students freedom to choose undefined pathways, you may begin to see the creativity that your students are capable of.

Adetoun Taiwo
Adetoun Taiwo is a Ph.D. student in the Department of Engineering Education at Virginia Tech. She has a Bachelor's in Biomedical Engineering and Master's in Mechanical and Nuclear Engineering from Virginia Commonwealth University. She has taught the second semester of foundations in engineering course for one semester and is motivated to find creative ways to keep engineering students engaged and interested while learning engineering. She sees engineering education as a tool for driving innovation and development in the world. She seeks to nurture interest in engineering through precollege initiatives. She wants to pursue research in the areas of innovation, engineering design and retention.

Creativity and innovation go hand-in-hand. It is fruitless to attempt to come up with new innovations that solve practical problems in our world without applying creative skill. Creativity is both individual and collaborative. The creative process is mysterious, intriguing, and elusive, and the demands of the twenty-first century call for more creative educational environments. Schools are called to nurture creativity and universities are called to become centers for creativity and innovation (Coate & Boulos, 2012).

Teachers and students alike can benefit from creative environments. In traditional lecture settings, there is a considerable amount of pressure on the student to "soak in" as much information as possible for later success on assessments like midterm and final exams. Even though there is a certain degree of necessity for a wide knowledge-base on particular core subjects, this system often does not foster the needed retention and transfer necessary for application of acquired knowledge in the future. This passive method of learning is especially difficult when there are barriers like fatigue or low attention span on the behalf of the students. Upon completion, a number of graduates leave their institutions with high grades but lack the ability to transfer their learning to new environments when they join the workforce. A curriculum which is fixed or compulsory may pose challenges to stimulating creativity—possibly more so than a curriculum which is more flexible (Craft, 2003).

The demand for skilled graduates, coupled with the increasing demand for creativity in the workforce, makes creative learning environments very crucial for students. Personally, I have been at both the giving and receiving ends of creative learning environments. As an instructor of a first year engineering class at Virginia Tech, Foundations of Engineering, I had the opportunity to apply creativity in my teaching. An important point to note is that the design of the course was done by the engineering education department and replicated in multiple sections, two of which I instructed. Most of the content, activities, and assignments were ready-made for me, but I had a little bit of freedom to make changes as I saw fit.

The semester began with an introduction to the design process, project-based learning, and a semester long project. Each student was placed in a small team to which they belonged for the rest of semester. I created the five-member teams, making sure that they were as representative as possible in terms of gender, skill level, and their access to resources. Over the course of the semester we covered topics and gave assignments that were related to their projects and helped them in preparing their deliverables.

Creativity was built into the design of this class because the structure of the class involved multiple modes with which students could learn. At the beginning of the semester, we did an activity called "the marshmallow challenge" to get students to learn about efficient problem-solving. The marshmallow challenge involves constructing a freestanding structure out of 1 yard of tape, 1 yard of string, 20 pieces of spaghetti, and 1 marshmallow in 18 min. The marshmallow is placed at the top of the structure and the tallest structure wins. We also watched a number of YouTube videos interspersed into the lectures over the semester. The students worked on a project and made prototypes in and outside of class. The content of meetings involved activities that applied creativity such as drawing, 3D modeling, writing, visual presentation, and oral presentation depending on the agenda for the day.

This same semester, I was enrolled in Practicum in Engineering Classroom (ENGE 5504) and it was a great opportunity for me to learn about teaching and apply my learning at the same time. My instructor for this class made our learning very engaging and interesting. The class was unique and it was evident that the instructor applied creativity in her teaching. In her class, we took turns reading poems at the beginning of each class and took a short period of time to respond to whatever poem was read. Some of the poems were directly related to teaching and some were not. We also did other things that seemed unrelated to teaching such as talking about our weekends, talking about the meaning of our names, and sharing stories of our diverse cultures. Something our instructor emphasized is that we can relate everything we were doing to teaching. In retrospect, each in-class activity either helped us interact or sparked our curiosity, which fosters learning as guiding dialogue and questioning to shape student learning (Collard & Looney, 2014).

Our class meetings were very interactive and we transitioned between multiple activities, keeping each meeting engaging. In addition to reading and responding to poems, other activities our instructor facilitated included regular discussions on assigned articles and book chapters that we read and our experiences with teaching on a weekly basis. Our classroom was also arranged in a way that we could all see each other, which helped our discussions. In addition, we took a short break about halfway through every class meeting. This allowed us to reset our minds and get re-energized for the second half of each meeting. In addition to discussing our experiences in class, we wrote reflections on our teaching experiences and would sometimes get feedback from our peers or the instructor depending on what we had written in our reflections.

The feedback from my practicum class helped shape some of the things I did in my own classroom. Over the course of the semester, I tried creative ways to keep my students engaged. Some of the class sessions involved learning how to use MATLAB. Prior to teaching this course, I had a low competency level using this programming software, and I was learning how to use it along with my students. In order to create an environment in which they could develop in their learning in spite of my level of competency, I gave them more opportunities than were required to practice using MATLAB in the classroom. I also followed a similar pattern when we started Computer Aided Design using Inventor. During those sessions, I walked around the classroom to see how they were doing and answered any questions they had. I also encouraged them to help each other, which reduced the burden on me to teach them everything they needed. They were more engaged and livelier during these sessions than they were during the sessions that involved me delivering content from lecture slides. During some of the practice sessions, I would play background music in the classroom.

For the first time, I had my students write a few points that they had learned, especially after I had taught them about some basic aeronautics. This gave me an opportunity to find out what they learned during the class meeting. Sometimes I saw the classroom as a laboratory where my students could try their hand at some activities and come to a new understanding, and other times I saw the classroom as a playground where they could have fun. In either case, I wanted my students to learn. There is a level of excitement that comes from being able to move around and interact in a classroom and I want to always have some degree of this component in my future teaching experiences. I have also come to value the importance of applying multiple modes of learning, as I believe this helps students connect with course material in a more personal and engaging way.

Last year I organized an event called "crafty circuits" in Ibadan, Nigeria for participants between the ages of 6 and 13 years old. The focus of the event was to inspire children in STEAM (science, technology, engineering, art, and mathematics) by introducing them to playful ways of learning. This event was modeled after "squishy circuits," an interactive model for teaching children about electricity using play dough made from food ingredients. We opened with an introduction using colorful and informative slides, and students did a hands-on activity. We took a break from the hands-on activity for lunch and offered a salsa dancing tutorial session. We then resumed the hands-on activity after the break and closed out. The participants had the opportunity to take their creations home as well as some of the supplies. In response to the question, "What do you hope to gain?" The participants wrote, "I want to learn new things"; "I hope to learn how to be able to do things big myself"; "I want to learn"; "I want to learn how to create"; "I want to learn how to be creative"; "I want to learn how to design"; "I want to gain a lot of things." From this sample of children, it is evident that we all want to learn creatively and creativity from a young age. This often changes over time as we grow older and get exposed to different rigid pedagogies and curricula, but to meet the complex demands of our emerging society, we need to be intentional about providing learning environments that nurture creativity (National Academy of Engineering, 2005).

Lindy Cranwell

Lindy is currently pursuing a part-time Ph.D. in Planning, Governance and Globalization (PGG) at Virginia Tech. The PGG program is administered out of the School of Public and International Affairs. Lindy's research will focus on faculty attitudes and meaning-making associated with comprehensive internationalization of university campuses and she has a passion for promoting professional and cultural skills for all university students.

Additionally, Lindy works full-time as Director of International & Graduate Education in the Via *Department of Civil & Environmental Engineering at Virginia Tech. In this position, Lindy works on global engagement, international programs, graduate program recruitment, and the overall running of the graduate program. Lindy also interacts closely with the College of Engineering, the Graduate School, and other on-campus committed to internationalization.*

"Transformative learning is grounded in approaches of risk-taking, exploration of self, collaboration and play" (Coate & Boulos, 2012, p. 130). This quote, found in an article about creativity in education, highlights the inherent creative quality of global education programs and experiences in the university learning process. As a full-time international educator in higher education and a part-time student (now working on a Ph.D.), I have lived this truth both as an administrator and student. A well-designed global education program has the power to enable students to be extraordinarily creative with their world and disciplinary views.

Ann Craft (2003) describes extraordinary creativity as "the production of new knowledge which has a major impact on an existing area of knowledge, the boundaries of which are monitored by experts within that field" (pp. 114–115). In my work, I have been part of planning faculty-led study abroad programs for several years. I help engineering faculty (experts in their field) organize meaningful learning experiences at international locations for undergraduate and graduate level students. One of my department's successful faculty-led international programs has been to the Dominican Republic. Faculty members led students to Punta Cana and Santo Domingo where students studied airport planning and sustainable water resources through a civil engineering lens. Prior to the classes/program, students were prepared in meetings for their upcoming learning experience. Students were taught about the Dominican culture, learned local dances, were introduced to the Spanish language, and were encouraged to think about the differences between engineering in the United States and the Dominican Republic. This type of program preparation was an individual and group exercise in risk-taking, culture awareness, group collaboration, and (definitely) play (the dance lessons were especially fun).

However, none of the preparation and self-discovery prior to the international experience compared to the learning experiences in-country. Engineering students were able to meet with local Dominican engineers and citizens, sharing hands-on work involving water sources and airport testing. Students attempted to navigate projects with limited data, unfamiliar environments, and a conflicting mindset about time and resources from the USA. One student expressed the richness of the experience by saying, "I loved being able to go out in the community to do hands-on work, rather than sitting in a classroom. Interacting with the locals and really making a difference in their community made this class hands down the best course I have taken" (Anonymous student, 2013). As students lived the differences in engineering between the USA and the D.R., they were transformed. Student learned to move out of their comfort zone and (I believe) never again approached civil engineering without considering a broader view of solutions.

In addition to my professional experience, I studied abroad as an undergraduate student. I can give a testimony to the strength of creative thinking generated by a degree program placed in a lived-experience setting. As an undergraduate in a Spanish language program, the cultural and language content in my courses was interesting, but the production of new knowledge during a six-week study abroad program to Mexico surpassed the acquisition of knowledge gained in multiple courses on campus. I believe that my positive experience in Mexico was not all by chance and location alone. The faculty leader was deliberate in the choice of courses, location, university, housing, and excursions to best provide a learning environment. She exemplified the "need for creative leadership within the university in order to face the challenges of the knowledge community" (Coate & Boulos, 2012, p. 131).

This past semester I was part of another example of deliberate creativity in teaching and learning. I was a volunteer teaching assistant for a professor in a freshman engineering classroom focused on teaching the concept of global culture and leadership in engineering. At the end of the course, all students (of this large class) participated in an international abroad trip to solidify the objectives of the class. Class activities and speakers were focused on providing the students with knowledge and/or practice with cultural differences and leadership both in society and the workplace. In one three-hour class session, communication specialists led the class through a simulation of different cultures. In this exercise, students were instructed on how to act within their assigned culture and had an opportunity to practice. Then the cultures had to interact and navigate communication without understanding each other. It was very uncomfortable for most students, but it brought to light what Collard and Looney (2014) said about disequilibrium's potential to "spur creative processes" (p. 350). This classroom was extremely student-centered with assignments designed to allow students freedom to explore and learn about the impact of engineering choices on society and vice versa.

Creativity in education has many definitions. Global education curriculum, programs, and experiences have inherent creative quality and serve as one avenue to motivate learning in university students. This particular type of creativity provides opportunities for students to take risks, explore who they are in relation to others, work in teams, and have a little fun. I have a passion for promoting international education. I look forward to future opportunities to use my passion for creative teaching in this regard. I believe that the opportunity to learn about other cultures, leadership styles, and approaches to academic disciplines *combined* with an international experience will propel students into transformative learning.

Rabih Younes

Rabih is currently a Ph.D. candidate in Computer Engineering at Virginia Tech. He has earned his Bachelor's and Master's degrees in Computer Engineering from the Lebanese American University, Lebanon. He has had a passion for teaching since he was young and started tutoring students when he was fourteen. Later, he taught technical engineering courses in colleges and universities. He is currently getting certified in Engineering Education at Virginia Tech and plans on becoming a professor after earning his Ph.D. For more information, you can visit his personal website: www.rabihyounes.com

Since the early 1900s, research has discussed how engineering education should be improved and how to apply the field's research practically within engineering classrooms (Mann, 1918). Today, just a few years away from 2020, engineering students are still very far from the vision that was discussed in the article "The Engineer of 2020" that was published in 2004 (National Academy of Engineering, 2004). In most universities, the same old curricula are being taught using the traditional methods of teaching that were in use decades ago without focusing on all the important advancements of engineering education research. Most people teaching in universities are adept researchers, but they are not always fit to teach twenty-first century students. Those faculty members may have never taken any courses, or attended any workshops or seminars about education in their whole lives. We even see new faculty, who have a deeper understanding of engineering education and are more up-to-date with current research in engineering education, also adopting traditional ways of teaching for a variety of reasons (Henderson & Dancy, 2007). James J. Duderstadt (2009), who was the President of the University of Michigan from 1988 to 1996, once said: "Most of our universities are attempting to produce 21st century engineers with a 20th century curriculum in 19th century institutions" (p. 4).

During my years in college and graduate school most of my classes, if not all of them, were based on lectures and techniques that go against most of the research that has been conducted in engineering education or in education in general. I have also taught college courses where I wanted to change the content and the way the class was offered, but I could not realize that due to multiple reasons. Those reasons include the heavy load of materials that are required to be taught class, making it very hard to incorporate more modern components, alongside departmental and higher administrative requirements, ABET requirements, etc. Those are some of the barriers that Henderson and Dancy (2007) discovered based on interviews with faculty members. These barriers include student attitudes toward school, expectations of content coverage, lack of instructor time, departmental norms, student resistance, class size and room layout, and time structure.

On the other hand, during my experience teaching Cisco Networking Academy (NetAcad) courses, which are very technical courses about computer networks, I have experienced successful methods of applying research to STEM educational practices, both at the instruction level and within the curriculum. The technical nature of these courses makes them a good candidate to use for comparison to technical engineering courses. The way NetAcad applied research in education to practice in teaching courses was mainly due to three updates to their old ways of teaching:

- Continually updating the curriculum to include all current changes and advancements in industry and academia.
- Continually updating the equipment used in classrooms and laboratories to fit current technology needs and advancements.
- Adding a requirement for instructors, which they called the "Instructor Trainer Qualification" (ITQ). The ITQ is a test that every instructor who plans on teaching a type of higher level classes (instructor training classes) should pass prior to teaching those classes. It mainly consists of a full-day test that has multiple parts. One part is about the instructor's mastery of course material, another part is about the technical skills of the instructor, and the rest is about their teaching skills and pedagogical methods (Cisco Systems, 2005).

We currently need serious research on how we could practically bridge the gap between the research in engineering education and how professors are currently teaching in engineering classrooms. Most methods that are discussed in previous research suggest major changes in the curriculum and reshaping nearly all the classes in the discipline. That kind of change takes a lot of time and is subject to a lot of resistance, whether from faculty or students (Henderson & Dancy, 2007). Previously proposed methods look very good on paper but they fail to make it to classrooms or are taking a lot of time to be integrated and applied. What is needed now, as an intermediate solution, are effective ways to apply current research that fits practically with the existence of current curricula and faculty. Something like NetAcad's ITQ should definitely exist to set minimum standards for engineering instructors. If a person is very good at conducting research, it does not necessarily mean that they would be a good teacher. Research and teaching are definitely correlated but being proficient in one does not necessarily indicate proficiency in the other.

Probably the most important issue in classrooms is that teachers fail to motivate students without the presence of exams. This type of extrinsic motivation makes students lose connection to the material after they take the exam. A heavy load of exams, within a culture of studying only to obtain good grades, transforms students into machines, forgetting about their dreams and creativity. Research exists on multiple ways to motivate students and let them be more interested and engaged. This research could be applied by instructors along with their administrative and curriculum standards.

On the other hand, not having enough time is one of the most significant obstacles standing in the way of applying engineering education research outcomes in classrooms. Programs are saturated with technical subjects that need to be covered by the end of the course in order for the student to be prepared for the next level.

Large amounts of required coursework presented in too-small units of time begin in early education and continue throughout a student's academic and professional life. This creates stress which affects the person's quality of life and damages their desire to become an engineer.

People often join engineering schools because they like to solve problems, they want to be able to accomplish some projects and dreams, and they want to help make the word a better place. Engineering students can end up forgetting about all those things and postponing them until an unreachable time in the future because of the way they are being taught in classrooms. Instructors might succeed in building a student who is knowledgeable about all the complicated theories in their field along with having some practical experience in the field, but they fail to foster the dreaming and creative part in the student. Therefore, on top of the urgent need for qualified teaching faculty, there is definitely a need to make some space in the curriculum for instructors to apply their special research-informed techniques in order to better fulfill their teaching responsibility.

Computers can now do a lot of things that instructors still spend a significant amount of time teaching in class. For example, calculus classes spend a lot of time on the subject of integrals, a subject so challenging that it could lead some students to hate the material and possibly their major. Nowadays, computers can perform these tasks more efficiently than any human. Students only need to learn the insight behind using integrals and how to use them to solve problems. For instance, students should learn about the multiple types of integrals and how integrals can serve them, then practice using them to solve real-life problems, and find their answers using computers. Making time in the curriculum and updating it to fit what the research in engineering education informs us is crucial to improve the teaching quality, and thus improving the quality of the graduating engineers.

To conclude, in order to create a twenty-first century engineer, we need teachers who follow and apply twenty-first century research that informs us about how creative ways of knowing in engineering would form better engineers.

References

Anonymous student. (2013, September 27). *Punta Cana 2013 post-trip assessment*. Retrieved from www.survey.vt.edu.

Bairaktarova, D., & Eodice, M. (2017). Thermodynamics in high energy rhymes and rhythms. *Advances in Engineering Education*. (*Forthcoming*)

Baxter Magolda, M., & King, P. (2012). *ASHE Higher Education Report* (Vol. 38, Issue. 3, pp. 1-138, Publication). Wiley Online Library. doi:10.1002/aehe.20003.

Cisco Systems. (2005). *Cisco Networking Academy Instructor Training Guidelines v2.0*. Retrieved from http://www.cisco.com/c/dam/en_us/training-events/netacad/academy_roadmap/docs/Instructor_Training_Guidelines.pdf

Coate, K., & Boulos, A. (2012). Creativity in education: Challenging the assumptions. *London Review of Education, 10*(2), 129–132. doi:10.1080/14748460.2012.691278.

Collard, P., & Looney, J. (2014). Nurturing creativity in education. *European Journal of Education, 49*(3), 348–364. doi:10.1111/ejed.12090.

Craft, A. (2003). The limits to creativity in education: Dilemmas for the educator. *British Journal of Educational Studies, 51*(2), 113–127. doi:10.1111/1467-8527.t01-1-00229.
Duderstadt, J. J. (2009). *Engineering for a changing world: A roadmap to the future of engineering practice, research, and education (Rep.)*. Ann Arbor, MI: The Millennium Project, The University of Michigan. Retrieved from http://milproj.dc.umich.edu/pdfs/2009/engineering%20for%20a%20changing%20world.pdf.
Henderson, C., & Dancy, M. H. (2007). Barriers to the use of research-based instructional strategies: The influence of both individual and situational characteristics. *Physical Review Physics Education Research 3*(2). Retrieved from http://journals.aps.org/prper/abstract/10.1103/PhysRevSTPER.3.020102
Jones, B. D. (2009). Motivating students to engage in learning: The music model of academic motivation. *International Journal of Teaching and Learning in Higher Education, 21*(2), 272–285. Retrieved from http://www.isetl.org/ijtlhe/pdf/IJTLHE774.pdf.
Mann, C. R. (1918). *A study of engineering education [the Carnegie foundation for the advancement of teaching (2010)]*. Boston: Merrymount Press. Retrieved from http://web.mit.edu/~jwk/www/docs/Mann%201918%20Study_of_Engineering_Educ.pdf.
National Academy of Engineering. (2004). The engineer of 2020: Visions of engineering in the new century. Retrieved from http://www.voced.edu.au/content/ngv:63792
National Academy of Engineering. (2005). *Educating the engineer of 2020: Adapting engineering education to the new century*. Washington, DC: National Academies Press.
Schön, D. A. (1983). *The reflective practitioner: How professionals think in action*. New York: Basic Books.

Afterword

The project of developing this collection sprang from a friendship and from a mutual respect for disciplinary expertise. After meeting Diana Bairaktarova and her students several years ago, I saw how introducing opportunities for creativity could enhance learning. I was impressed with the poems and videos students produced to depict the laws of thermodynamics, but I also realized how challenging it was to move students to try the creative approach. I am referring to a wonderful group of young people who, it seemed, had already put away the creative drive that likely influenced their path to engineering in the first place. Fortunately, Diana re-awakened that drive with her energy and assignments.

I am thankful that in my lifetime perceptions of creativity have moved from seeing it reside in exceptional individuals who are somehow set apart from the rest of us to understanding the potential for creative opportunities in everyday learning and teaching. Like other affordances for learning, these context-sensitive opportunities must be thoughtfully designed for creativity to flourish.

When we orient students to the problem-based thinking foundational to engineering, what some of us fail to realize is that the creative arts are also problem-based in that the intersection of what is and what might be longs to be revealed in some form. Bridging the STEM fields with the liberal and creative arts is not a new assembly; as Diana Bairaktarova notes in her introductory chapter, that synergy is historically found in the scientist-artist identity. The question becomes how do we reinstate this synergy into the current curriculum of higher education? It seems our colleges of engineering are often big islands in the middle of our campuses; the challenge is to build partnerships that bring excitement about creative ways of knowing.

All fields are converging around issues and problems urgent to our planet. The required innovative approaches can be leveraged with creative ways of knowing: "To creatively solve open-ended problems that have no known solutions involves divergent or intuitive thinking skills" (Feist, p. 287). Setting up the conditions for learning in these ways is what we hope this book helps readers to do.

D. Bairaktarova, M. Eodice (eds.), *Creative Ways of Knowing in Engineering*,
DOI 10.1007/978-3-319-49352-7

Finally, a collection like this is an act of "collective virtuosity"—a term used by Diane Dabby in this volume to describe the improved performance of a group based on the work the group does together. We offer engineering education an example of how collaboration with other disciplines can lead to a future of creative solutions.

Reference

Feist, G. J. (1999). The influence of personality on artistic and scientific creativity. In Sternberg, R. J. (Ed.), *Handbook of creativity*. Cambridge: Cambridge University Press.

MIX
Papier aus verantwortungsvollen Quellen
Paper from responsible sources
FSC® C105338

If you have any concerns about our products,
you can contact us on
ProductSafety@springernature.com

In case Publisher is established outside the EU,
the EU authorized representative is:
Springer Nature Customer Service Center GmbH
Europaplatz 3, 69115 Heidelberg, Germany

Printed by Libri Plureos GmbH
in Hamburg, Germany